Introducing X.400

P A Chilton

PUBLISHED BY NCC PUBLICATIONS

British Library Cataloguing in Publication Data

Chilton, Paul
   Introducing X.400
   1. Digital communication systems
   I. Title
   621.38'0413

   ISBN 0-85012-692-4

First published in 1989 by:

NCC Publications, The National Computing Centre Limited, Oxford Road, Manchester M1 7ED, England.

Typeset in 11pt Times Roman by H&H Graphics, Blackburn; and printed by Hobbs the Printers of Southampton.

ISBN 0-85012-692-4

Acknowledgements

I am grateful to the following people who kindly offered to review material for this book and provided valuable comments:

Janusz Zajaczkowski of British Telecom

Bill McKinley of British Airways

Bob Willmott, an Independent Consultant

In addition the author wishes to acknowledge the CCITT X.400 red book recommendations and the draft CCITT X.400 blue book recommendations as a major source of material for this publication.

The Centre acknowledges with thanks the support provided by the Computing Software and Communications Committee (CSC) for the project from which this publication derives.

Contents

Appendix

1 Introduction

GENERAL

Within the world of business and commerce the need for the rapid transfer of information has always existed. Information in its various forms is a valuable commodity which, if acquired in advance of competitors, can provide a business with its competitive edge. Evidence of the importance of information can be gained by the rapid growth of companies whose business is purely based on information gathering and distribution, whether via paper-based (directories and circulars) or electronic (database) means. Examples here would include such companies as Dunn & Bradstreet, Reuters, Kemps, Finsbury Data Services, Pergamon, etc.

Information is not necessarily gathered from specific sources, such as those above; it may originate from many different areas and via numerous media. Within large companies a vast amount of information is generated and this will also need coordination and distribution.

Although, for some considerable time, postal services and paper-based documentation have constituted the backbone of information transfer, it is now obvious that electronic means are having a growing impact. The introduction of telex, in 1932, was an important early step and its current usage and penetration is impressive. However, within the last decade, there has been a gradual introduction of computer-based office automation equipment. The range of facilities covered by office automation is very broad, but in the context of information retrieval and transfer it can be narrowed to encompass the following technologies:

— telex;

— teletex;

— public electronic mail;

— private electronic mail;

— facsimile.

At the same time the computer is increasingly penetrating the business environment. Initially the computer was used to achieve very specific objectives or tasks. This meant that a computer would be the property of a single department or group and that it would serve particular needs. Examples here might be payroll calculation or other general data processing applications. Information storage, a major use of the computer, is exploited by the various applications.

The development of the computer since its early days has been very rapid. Today it is possible to buy a personal computer which has the memory storage capability and processing power which just over a decade ago might only have been available in mini and mainframe installations. This rapid development has led to the distribution of computer processing power throughout the office, even down to the level of individual desks.

Experience with computers in particular and office automation (OA) equipment in general has yielded a user requirement for increased levels of interconnection. In this way information and services, formerly only available in small pockets, can now be distributed over the office or indeed an entire organisation. The sort of communications required to support this activity have been slow to develop for various reasons:

— product manufacturers wishing to protect their existing user base and hence not producing links to other vendors' equipment;

— lack of internationally developed and agreed interface solutions;

— lack of coordination of the user requirement for such developments.

Individual vendors have specified product architectures which allow the interconnection of their own equipment, but connection between different vendors' equipment has always been a problem because of the lack of an agreed interface. What is required is an internationally standardised interface solution which will allow differing technologies to communicate using hardware purchased from different suppliers.

This ultimate goal is not now some distant oasis on the horizon. The standardised interface solution has arrived under the guise of the International Standard Organisation's (ISO's) work on Open Systems Interconnection (OSI). The OSI framework and its internal operations define a standard interface for the reliable transfer of data between computer systems. Architecturally this work is based around a seven-layer model which effectively separates the functions into logical groups.

Within the OSI framework the actual user applications (ie the functions with which the user will interact) reside at the uppermost layer. X.400 message handling, as defined in the CCITT's X.400 recommendations, is amongst the first OSI-based user applications to have been defined so far. X.400 defines a method for carrying out medium independent electronic messaging, on a store-and-forward basis, between computer-based systems. The term 'store-and-forward' should be noted as it means that these systems are only suitable for applications which are not real time, ie those that do not require an instantaneous question-and-answer type working. With that proviso, however, these systems can act as a medium for transporting any form of structured or unstructured data. In this way they may act as a platform for many different industry specific applications. Because X.400 offers a standardised solution, it can also be used to access the integration and interconnection of the information transfer technologies which have evolved in today's business environment.

Because the X.400 recommendations are internationally defined and agreed, it means that companies all over the world will be able to build conforming products secure in the knowledge that they will be able to interwork with other X.400 vendor products. It is likely that, during the next couple of years, X.400 Message Handling Systems will develop to become the most important part of the future worldwide messaging infrastructure. It will also be instrumental in aiding the worldwide adoption of ISO OSI communication standards. This is the only way in which users will achieve the equipment vendor independence which they all crave.

(Terminology Note: The CCITT X.400 recommendations are a series of eight closely related documents which are numbered in the range of X.400 to X.430. Although X.400 is one of the individual recommendations within the series it is acceptable to use the term X.400 when referring to the series of recommendations as a whole. This convention is used throughout the remainder of this book.)

PURPOSE AND SCOPE OF THE BOOK

The aim of this publication is to give the reader a 'first-level' technical introduction to the subject of Message Handling Systems (MHSs) based on the X.400 recommendations. Chapter 2 looks at the general area of electronic information transfer by introducing the relevant technologies. This base information will give the reader sufficient insight into the subject to be able to consider the potential advantages of the X.400 approach.

Chapter 3 provides an overview of the principles of Open Systems Interconnection (OSI) upon which X.400 systems are based. It looks at the advantages of this approach and then proceeds to outline the functions of the seven-layer OSI reference model. Building on this knowledge base, Chapter 4 looks at the technical concepts of X.400 MHSs by identifying the functional elements and the services which are provided by them.

The most important aspect of any implementation of information technology is to establish what exactly will be the benefits. Chapter 5 outlines some of those benefits which can be achieved by adopting X.400 systems. It is only by considering these that it is possible to assess the impact of this technology on any specific organisation.

The CCITT have sustained their effort on the topic of X.400 since the ratification of the original recommendations in 1984. This has culminated in the development of completely new documents which, while retaining backward compatibility, have been extended and improved in many respects. Chapter 6 looks at these recent developments.

It must be emphasised again that this book will only yield a first-level introduction to the topic of X.400. Further study is essential when considering implementation, and it would be advisable to seek as many information sources as possible. X.400 implementation is a complex subject which should not be taken lightly, but this should not discourage an organisation which is looking towards long-term technology benefits.

2 The Need for Standardised Electronic Messaging

INTRODUCTION

Means of electronically transferring human-readable information, between geographically separate locations, have existed for many years. Most people will have witnessed the operation of one of the earliest forms of electronic messaging in the classic 'wild-west' movies. The telegraph system, which always appeared threatened by wire-cutting activities, was probably the first commercially available electrical/electronic messaging system. Here the message to be sent was coded into electrical pulses by a specially trained operator at the point of transfer. The coded information was then carried along special lines (provided they were intact!) and decoded at the receiving station by a similarly trained operator. The message was written down as it was decoded, and the resultant paper-based message was then delivered by hand over the short distance to the intended recipient.

The next major development in this non-voice mode of inter-human communications was telex, introduced in 1932 and still with us today. Telex had the major advantage of being a direct facility between the originators' and recipients' premises, hence removing the need for the sometimes unreliable human delivery mechanism. In addition no special skills were required to operate a telex terminal apart from those (eg typing) of the normal office worker.

Telex remained the major form of inter-business communication for many years until rapid developments — indeed, quantum leaps — were made in the electronics components industry. The discovery of the electrical properties of semiconductor materials such as silicon and germanium led to the development of the transistor as a control device with the same operational characteristics as the electrical valve, but with greatly reduced production costs. Other devices such as diodes, FETs, thyristors and triacs were also developed. As experience grew with the new materials it became possible to influence many of the operational characteristics of the devices made from them. Alteration of the impurity doping levels of the base material allowed manufacturers to dictate the performance of their devices in terms of power handling capacity and speed of operation.

The next major electronics progression was the Large Scale Integration (LSI) of devices. This allowed more and more devices to be integrated on a single piece of semiconductor, usually silicon because of its better characteristics. Prior to this all the computers which had been developed consisted of huge numbers of discrete components, ie valves or transistors. This meant that they were truly massive machines (generally filling reasonably sized rooms) which lacked reliability and commanded purchase prices that related to their immense physical size. LSI and subsequently Very Large Scale Integration (VLSI) offered the potential to produce large numbers of interconnected devices, hundreds and even thousands, on a single small piece of silicon. Such developments would obviously have a dramatic impact on the size, practicality, cost and performance of the computer.

The computer is not a random mixture of switches formed by its transistor elements; it has specific units for specific tasks. A good example is the Arithmetic and Logic Unit (ALU), concerned with arithmetic and logical operations on the data contained within its memory banks. Instead of implementing all of these units in separate blocks and then making connections via external wiring, it seemed sensible to integrate some of these together upon a single device. The microprocessor, a realisation of this idea, is basically a computer on its own. Its memory banks must be supplied on additional devices, although it can address the data elements on these additional devices. The microprocessor has revolutionised the electronics industry and has been instrumental in providing today's wide range of computer-based alternatives to telex. Within this current list of options for the user of electronic messaging are the following systems and services:

— telex;

— teletex;

— public electronic mail;

— private electronic mail;

— facsimile.

All of the above list, including telex, have been affected by the development of the microprocessor and its employment in microcomputers, minicomputers and specialist terminal equipment. While the microprocessor has *helped* the development of the telex service, particularly its terminal equipment, the microprocessor was *essential* for the development of the other items in the list.

In addition to the developments in computer hardware, there is a crucial legislative development which has served to enlarge the user's choice of communication systems and services. The liberalisation of telecommunications, on an international basis, has enabled companies to provide third-party value-added services, such as public electronic mail. These services make use of existing PTT circuits by 'adding value' in terms of additional facilities, such as storage, management and access to databases. Previously PTTs had operated in monopolistic situations without the drive of fierce competition for the provision of services. Deregulation of the telecommunications environment has led to more imaginative usage of existing services and a greater awareness of the needs of the customer.

The major problem with today's wide choice of systems and services, however, is that there is little or no interconnection between them because of the lack of technology compatibilty.This means that the individual user bases of each of these services/systems remain separated, to some extent, by their own technology boundaries. This problem was one of the major driving forces behind the development of the CCITT's X.400 recommendations which effectively provide a standardised interface. But before the merits of this approach can be considered it is important to examine the existing alternatives. In this way their strengths and weaknesses can be assessed, and the role to be assumed by X.400 systems will become clear.

TELEX

Telex is a service familiar to all organisations and individuals. It offers its users a facility for the interchange of simple text-based messages on a point-to-point basis. It is essentially a slow-speed service which lacks functions and facilities when compared to some recently developed forms of electronic messaging. Recent improvements in the design of terminals, however, have meant that the service has maintained (and indeed in some cases increased) its popularity. Many organisations still rely on telex as their main means of communication in the business environment, and hence it has become crucially important for their livelihood

So why is it so popular? Firstly, it is an international service which has a worldwide user base of around 1.5 million spread over 200 countries. Secondly, it is still the only telecommunications service which provides a hard copy, at both ends of the link, which is internationally recognised as a legal document. This legal status would seem, however, to be debatable: British Telecom have recently stated that they are not aware of a precedent, on which English law is largely based, establishing a telex message as a legally binding document.

Most of the more recent developments in text communications, such as electronic mail (public and private) and teletex, have offered a gateway through to telex (though in the case of teletex this is not now available, along with other Interstream services) because of its large installed user base. Access to this existing base is an essential factor in providing the critical mass of users so essential to get a new service off the ground.

Some of the traditional drawbacks to telex have been addressed by the arrival of computer-based systems. By means of a personal computer or mini/mainframe computer workstation, a telex can be received and read, or written and transmitted, without the users having to leave their desk. Moreover, many computer-based telex packages take care of sending and receiving messages in background memory so that the users may carry on with their current task without having to wait for, or to attend to, the receipt of a message.

The advantages of telex are that:

— it is easy to use;

— it offers a cheap service;

— it offers a large worldwide user base;

— it offers interface capability with other networks.

The disadvantages are that:

— message preparation facilities can be poor;

— it is prone to error;

— messages are not secure;

— it has a limited character set;

— it has slow speed;

— it is location dependent;

— it is resource dependent.

Its primary application is in the delivery of short text messages.

TELETEX

Teletex is a service which adheres to an internationally agreed standard for the electronic transfer of documents. It provides for the high-speed transmission and receipt of text-based messages on a point-to-point basis. It can be viewed as a sort of super telex service offering its users improvements in terms of transmission speed, accuracy of messages, enhanced message preparation facilities and better-quality hard copies.

The teletex service enables users to prepare text documents on teletex compatible devices such as electronic typewriters or word processors, and then transmit them via a public network to other, maybe dissimilar, devices having teletex compatibility or via a gateway to the national and international telex services. Local mode operation (text preparation, editing, printing, etc) is not normally affected by the transmission or receipt of a document, as this is a memory-to-memory communication system.

The teletex service was conceived by the German Bundespost in the mid-1970s and was developed in co-operation with Siemens. The development was due to two major problems which faced Bundespost:

— linking of dissimilar text systems, with the associated problems of incompatible character sets, protocols and formatting controls;

— the outdated telex service. Telex is slow, prone to error, has poor presentation with equipment which lacks a good user interface and text creation facilities. This latter problem has been eased slightly with modern telex terminals.

It was from these beginnings that teletex evolved and was subse-

quently taken on board by the CCITT as an international standard for
text communication.

For users who have word processing facilities and require access to
this service there are a number of options depending on the type of
system that is available.

— **Teletex terminal** This does not have to be a dedicated terminal;
it can range from a simple electronic memory terminal to a word
processor, business computer or combined workstation.

— **Adaptor or 'black box'** This can add teletex capability to an
existing word processor (much in the same way as a telex
adaptor), sophisticated telex terminal or computer by means of
a software package.

— **Multi-user** Where more than one user on the system needs
access to teletex some means of distributing the facility must be
found. This is usually via a cluster controller, Local Area
Network (LAN), or PABX.

Although teletex was initially devised as a replacement for telex , this
has not occurred in any country except Germany (in excess of 17,000
terminals). In the UK, BT launched their service in 1985 but it has failed
to capture user interest (less than 300 registered users in the UK) and
now they are to remove the Interstream telex gateway facility (mid-
1988).

The advantages of teletex are:

— internationally standardised service;

— in-built error detection;

— high speed service;

— good message preparation facilities;

— messages received when equipment is unattended/'no power'
receipt;

— extended character set;

— document not message-based.

The disadvantages are:

— small user base in the UK;

— can be expensive to get started;

— document not message based;

— location dependent.

Its primary application is for the fast delivery of text-based documents.

PUBLIC ELECTRONIC MAIL

An electronic mail (E-mail) service will provide its users with person-to-person communication service for text-based messages. When you register with an E-mail service you are allocated a password-protected mailbox or message storage area on the system's host computer. This means that, if you are unavailable, any messages will be stored until you can access them at a later time or date. Only people who know the password for a particular mailbox can gain entry; hence, the systems are quite secure. E-mail services also provide their users with access to on-line databases, noticeboard facilities, the telex service and other useful features.

These services, such as Mercury Link 7500 (was Easylink), Quik-Comm (Geisco), Comet (Istel) and One-To-One, have been allowed to evolve in the UK since the 1981 Telecommunications Act. This enables organisations to provide services over the public switched telephone network (PSTN) and the packet switched service (PSS), whether or not they compete with the services which British Telecom (BT) offer (in this case, Telecom Gold).

To gain access to these systems requires that the user is registered with the service and has a computer with the appropriate software package. The user may utilise a micro or be connected to a multi-user mini/mainframe installation. A communications software package is required for the micro/system in order to allow it to download messages, files and information from the E-mail service computer and for uploading text prepared off-line; the package may be tailored to the particular service or be able to be used in any environment. An approved modem, acoustic coupler, Dataline connection to the local PSS packet switching exchange (PSE) or a Multistream service can be used for dial-up access to the E-mail service.

Although such services offer reasonable features and facilities, they seem to fail to meet user criteria in the area of interconnection. Such systems and their associated user bases have remained unable to intercommunicate between themselves since their introduction. This situation seems, not unnaturally, to be unreasonable as computer systems generally seem to be moving towards higher levels of integration.

The advantages of public electronic mail systems are:

— low starting costs, especially if a micro is available;

— access to other value-added services, such as telex, databases, etc;

— contacting people who spend time out of the office, transferring text via portable PCs and acoustic couplers;

— the message security offered by the password-protected mailboxes;

— reasonably large user bases.

The disadvantages are:

— high running costs;

— no connection between different vendors' systems.

Public electronic mail services are mainly used for fast, location-independent delivery of messages rather than extensive documents.

PRIVATE ELECTRONIC MAIL

The provision of electronic mail facilities within private organisations is now an established means of achieving fast and efficient intra-business communications. Such private electronic mail systems will offer either similar or more advanced features than those of the public electronic mail services, but are provided on corporate hardware. In the past, organisations have been forced into the purchase of incompatible systems which has led to special difficulties when interconnection is required. The task of integrating a number of corporate electronic mail systems, from different sources, is not easy. In such cases it is difficult to achieve acceptable results and can often lead to considerable expense, due to the one-off nature of such problems.

The options for an organisation wishing to set up its own internal electronic mail service fall into one of two categories:

— electronic mail facilities provided as part of an integrated office system such as DEC's All-In-One or Data General's Comprehensive Electronic Office (CEO);

— licensed electronic mail software from an electronic mail vendor (an example here would be Istel's COMET which is available in two forms, one to run on DEC systems and another version for IBM systems). Such packages may be customised to cope with individual customer requirements (an example here might be the integration of the mail system with existing word processing facilities).

Integrated office systems attempt to integrate all the logical office facilities within one system, such as:

— electronic mail;

— word processing;

— graphics;

— spreadsheet;

— diary;

— communication to outside services such as telex, prestel, etc;

— document filing and retrieval.

Such systems are extremely powerful, within their own environment, and will allow the integration of text, data and graphics within documents which can then be transferred between users.

The advantages of private electronic mail systems are:

— minimal running costs;

— good security, reliability and availability;

— possible to access other value-added services;

— location independent;

— can be message or document based.

The disadvantages are:

— high initial costs;

— in-house support required;

— lack of interconnection with external systems and those from
other vendors;

— lack of impact.

Private electronic mail systems are mainly used for intra-organisation
document or message communications.

FACSIMILE

Facsimile is a service which can transmit and receive pre-prepared text
and graphic information. The process employed is straightforward; in
simple terms, a document is photocopied and the image is encoded on
the paper as a series of black or white elements. A light-sensitive
scanner reads the original document, dividing the image into thousands
of squares. This scanned image data is then converted into digital
impulses which are in turn converted into analogue signals, via a
modem, suitable for transmission over the public telephone network
(national or international). The receiving machine will interpret the
signal and re-create the original image.

In the early days, facsimile machines were slow and generally each
manufacturer produced their own incompatible implementations. The
speed of the machines has been significantly improved with the recent
rapid developments in electronics hardware but the problem of incom-
patibility had to be tackled by the adoption of international standards.
The CCITT (International Consultative Committee for Telephone and
Telegraph) some 15 years ago developed the first international standard, Group 1, and since then the market for facsimile has grown
rapidly. Today standards for Groups 1 to 4 are established.

All facsimile machines are now manufactured according to the design
standards agreed by the CCITT. Terminals have been divided into the
four groups:

— Group 1, established in 1968, used analogue transmission
techniques and took nearly six minutes to send an A4 page.
These machines are now virtually obsolete.

— Group 2, established in 1976, also uses analogue transmission
techniques but takes three minutes to send an A4 page and has
a better image quality at the receiving end. These machines too
are rapidly being replaced by, or passed over for, Group 3
machines.

— Group 3, established in 1981, is now the worldwide machine standard, especially for international communications. By utilising data compression techniques these machines make the best possible use of the transmission medium, and hence achieve fast document transmission times. A4 document transmission time has been reduced to less than a minute with the development of Group 3.

— Group 4, established in 1984, allows facsimile transmission to take account of the developing public digital telephone networks. When a fully interconnected digital network has been established the use of Group 4 machines will become viable and should allow very rapid transmission of documents. Transmission times for an A4 document have been quoted as low as five seconds via Group 4 technology.

The emergence of these standards has had a massive impact on the user adoption of facsimile transmission. In only a very short period its use has escalated to the point where it now rivals telex as the most popular means for document transmission.

In the past the only method of sending a facsimile was to produce or obtain the required text and/or graphics in a hard-copy form and then go to the facsimile machine and send the document. This generally involves a person having to leave their desk or place of work and then having to spend time establishing the phone connection before being able to send the information. This problem is now being addressed by the arrival of personal computer PC-to-fax links.

Because of the rapid proliferation of facsimile terminals and personal computers in the office environment, it was only a matter of time before someone devised a method of exploiting the attributes of both devices. A PC-to-fax link consists of a software package and a modem for the PC which allows you to digitise the information on the screen into the required format for a fax transmission. The modem is then used to send the information via the PSTN to a dedicated fax terminal or another PC with a fax link. The software can handle the task of sending the information as a background task and hence the sender is free to continue with another task. Hence with a PC, the required software, a modem and a phone link, a person will be able to send messages without ever having to leave their desk.

The advantages of facsimile transmission are that it:

— can transmit both image and text information separately and together in the same document;

— is easy to use;

— requires no special training for users;

— allows higher speed transmission to become commonplace;

— facilitates low transmission costs.

The disadvantages are that it:

— is not secure;

— is prone to errors;

— does not integrate into existing systems (PC-to-fax links will help);

— is location dependent;

— is resource dependent.

Its primary application is in the fast delivery of mixed text and graphics documents of varying length.

PROBLEMS WITH ELECTRONIC MESSAGING SYSTEMS

In each of the sections on the individual electronic messaging technologies, a summary of the advantages and disadvantages was provided. Each technology offers a reasonable service to the user, with the main disadvantages surrounding the areas of cost, lack of facilities, small user base, interconnection and integration.

The cost of services/systems has always been a source of contention between suppliers and users. The cost of computer-based hardware has reduced drastically with the emergence of LSI and VLSI, and this in turn has yielded scope for improving the facilities offered by such systems. Because of this, the cost of hardware has appeared to remain at the same level while in real terms this represents a reduction: newer products offer much increased performance for the same cost. The costs of computer-based messaging services may appear high but almost on a daily basis one or other of the suppliers is providing additional functionality (or further 'added-value'). It is this added value (eg access to external databases, notice boards, etc) which can sometimes be crucial in justifying the expenditure on this type of service. Although the points mentioned here would seem to offer some mitigation for the first two disadvantages, quoted at the beginning of this section, the later two have yet to be successfully addressed by any system or service.

The lack of interconnection abilities and/or failure to integrate with other OA equipment have yet to be addressed by all means of electronic information transfer except telex. Telex, however, although it seems to integrate with most systems, except facsimile, has its own problems: it lacks facilities/functionality and speed when compared to computer-based systems. Lack of interconnection — between public electronic mail services, for example — means that they are instantly restricted in terms of the potential user base which can be addressed. In the UK there are more than four services available, none of which will talk to each other. The integration of differing technologies is the only way in which their full individual potential for improving communications and productivity can be realised. Facsimile is particularly guilty in this area, mainly as it is based upon the transfer of scanned, rather than the more common text-based, information. This has meant that until recently, with the arrival of PC–fax links, facsimile documents have existed in paper-based format only. This implied a manual transformation of the information from this format (ie re-keying) before it could be stored or manipulated electronically.

WHAT DOES X.400 OFFER?

The CCITT's X.400 Message Handling System (MHS) recommendations define the generic architecture and underlying protocols for an all-purpose message handling service. These recommendations are internationally agreed. Hence they provide a common basis from which the developers of office systems and electronic mail services can produce systems which can be interconnected, and hence interwork.

The system defined in these recommendations is basically a store-and-forward electronic messaging service which has the ability to transfer any type of information, whether it be text, image or pure data. *Store-and-forward* means that such systems are not real-time: the time when a message is offered to such a service for transport bears no fixed relationship to when the message is actually delivered to the intended recipient or system. X.400 systems are effectively separated between the user applications and the underlying means of message transfer.

This feature allows multiple user applications to be run over the same link. Initially at least the major application will be interpersonal messaging. This will operate in a similar manner to existing public electronic mail services, allowing the interchange of messages and documents between human users. Ultimately, however, multiple user applications (eg E-mail, EDI and processable document exchange) will all be able to operate over the same basic transport mechanism.

Because these recommendations are internationally agreed and available in the public domain, it means that X.400 can serve as the basis upon which a worldwide messaging infrastructure will be established. Existing messaging systems and services can be interconnected via the use of X.400 protocols, which in practice will mean:

— service provider or PTT provision of connections to other X.400 systems around the globe with access units from their X.400 based services to telex, facsimile and teletex;

— equipment-vendor provision of X.400 gateways or complete systems based on X.400 which will enable the interconnection of a private system with another or with a public message handling service.

The X.400 solution is future proof because any development of the current recommendations will maintain compatibility, hence companies will be no longer confronted with the dilemma of what to buy and when to buy it. Users will be able to enjoy the vendor independence which they all crave. Also because all vendor products will conform to the same standard they can be selected purely on merit rather than according to their ability to interconnect or integrate with existing product purchase decisions.

It was stated earlier that X.400 systems can act as a platform for a wide range of applications, other than just electronic mail. This really is the key to the usefulness of such systems because applications can be updated, replaced or added as is deemed necessary. In this way an X.400 system can grow in the future to encompass applications which have not yet even been considered. This means that when considering such a system it can be looked upon as an all-purpose data transfer mechanism which has enormous potential for future development.

It should by now be apparent that the adoption of systems based on X.400 has some substantial benefits when compared to existing means of electronic information transfer. These include:

— internationally standardised solution;

— future proof;

— a basis for industry-specific applications, not just electronic mail;

— interconnection and integration of existing technologies.

The benefits of X.400 are expanded in Chapter 5, along with an indication of potential user applications for services based on the recommendations.

3 Open Systems Interconnection (OSI)

INTRODUCTION

The CCITT work on the X.400 recommendations has evolved from the field of international communication standardisation and, more especially, from the topic area of Open Systems Interconnection (OSI). This chapter introduces both these areas, effectively 'setting-the-scene' for the following explanation (in Chapter 4) of the technical concepts involved in the X.400 Message Handling System (MHS) recommendations.

INTERNATIONAL STANDARDISATION ORGANISATIONS

Looking into the world of international communication standardisation, especially for the first time, can be a confusing and fruitless task. There is a wide range of organisations operating from various countries which sometimes appear to be working in the same areas and hence duplicating work. The names of these organisations, often long and unmanageable, are shortened to acronyms which serve only to confuse the newcomer.

This initial shock of finding such confusion in computer communications does not encourage faith in the belief that anything fruitful will ever emerge.

This section introduces some major standardisation organisations and indicates how they co-operate. In this way the reader will gain some insight into the process of international standardisation.

MAIN STANDARDS-MAKING BODIES

The major bodies involved in the standards-making process are introduced and details are given of their current work on International

Communication Standards. Figure 3.1 shows the relationship between
these groups and other contributory bodies.

International Standards Organisation (ISO)

ISO is an agency of the United Nations (UN) which was established in
1946 with voluntary member bodies, currently over 70, who can be
either participating (voting) or observing (non-voting). ISO has the
prime (but not exclusive) responsibility for international standardisation. Two important stages in the progression of a standard through the
ISO machinery are: 'draft proposal' (DP) — when a standard has
passed the formative stages it is given the number (eg DP 9999) under
which it should eventually be published; and 'draft International Standard' (DIS) — when a draft has been accepted, by a vote of the committee responsible for producing it, the DIS document is put forward for a
wider international vote by the members of ISO, the national standards
bodies. ISO standards are usually subsequently approved by these
bodies and reprinted as national standards.

Comité Consultatif International de Télégraphie et Téléphonie (CCITT)

The CCITT is part of the International Telecommunications Union
(ITU) and is a treaty organisation whose members, mainly the service
providers and PTTs (ie Post, Telegraph and Telephone authorities) of
member countries, have to sign a convention to join. The CCITT major
role is the harmonisation of communications across the world. Every
four years, CCITT publishes a series of recommendations on various
communication standardisation topics (one of which is X.400); each of
these supersedes any previous issues on the same topic. These recommendations are binding on the member countries in the realm of
international communications.

International Electric Commission (IEC)

The IEC (established in 1906) has a voluntary membership of national
elements called 'national committees'. In the past there were areas of
overlap between ISO and IEC but these have now been eliminated, and
the IEC now concentrates on standards addressing product safety and
environment.

American National Standards Institute (ANSI)

ANSI is the US agent and voting member of ISO. It is a private and voluntary organisation supported by its membership, which includes representatives of manufacturers, research groups, standards groups and other interested paying parties. The organisation is supported by membership dues and document sales.

National Bureau of Standards (NBS)

The NBS is an agency within the US Department of Commerce and is funded by the Federal Government to develop standards. NBS has responsibility to act as an investigatory, formative, approval and publishing organisation for standards in Federal Government. Despite its narrow target of the Federal Government, the combination of all the main roles (backed up by a very substantial budget) means that it can exercise a degree of project management which is impracticable where responsibilities are split. As a result it is highly influential in the much wider circle of national and international computing standards. NBS has strong links with other national organisations such as ANSI and IEEE.

EIA and IEEE

Electronic Industries Association (EIA) and the Institute of Electrical and Electronic Engineers (IEEE) are professional societies in the US. Both of these organisations have contributed to the work on OSI particularly with standards for the lower layers of the reference model.

European Computer Manufacturers Association (ECMA)

ECMA is a group which consists of about twenty major European manufacturers. It concerns itself primarily with computing standards, but carries out none of the other functions of a Trade Association. Its committees are open to users as observers, but only its member manufacturers have a vote at the approval stage. ECMA serves mainly as a non-voting contributor to both ISO and the CCITT and intends to support the protocols adopted by ISO for OSI.

Commission for the European Communities (CEC)

The commission acts as the civil service for the European Economic

Community (EEC). The Commissioner for Industrial Affairs is responsible for telecommunications and information technology policy.

La Conférence Européene des Administrations des Postes et des Telécommunications (CEPT)

The CEPT is a conference of the posts and telecommunications authorities (PTTs) for Western Europe and certain European economic organisations.

British Standards Institute (BSI)

BSI is the UK national standards body who make contributions to ISO work and have a voting status, as the UK representative. They also publish ratified ISO standards

Association Française de Normalisation (AFNOR)

AFNOR is the French equivalent of the BSI in the UK.

LIAISON BETWEEN ORGANISATIONS

There is considerable overlap in the work carried out in European standards organisations. Efforts have been made to minimise this in ISO and CCITT groups working in the same areas: regular liaison meetings are held to ensure that work is progressing in the direction that is most constructive for everyone concerned.

Figure 3.1 indicates the general structure of the world standardisation process. The groups like ECMA and other industrial committees, along with national standards bodies (such as BSI and ANSI), contribute to the actual world standards-making process that revolves around ISO and CCITT.

The groups indicated above have all been involved in the formulation of the standards for Open Systems Interconnection (OSI). Obviously ISO, with their responsibility for International Standards (IS), have been committed to the OSI work for some time. CCITT have also been heavily involved with work on OSI standards and have ratified many recommendations which have since been taken on board by ISO for progression to full IS status. This close cooperation between the groups reflects the scale of the tasks to be accomplished by the OSI work.

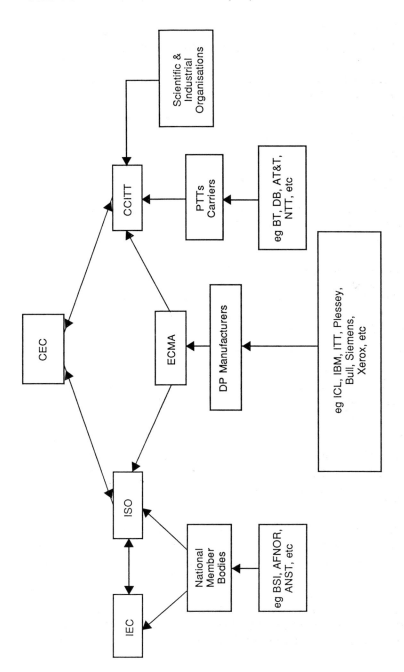

Figure 3.1   Interactions Between Standards Bodies

CONCEPTS OF OSI

The interconnection/interworking of office automation and other computer-based equipment has been a problem for the user for some time. In the short term proprietary communication and equipment standards have gone some way, within single vendor domains, towards alleviating the problems. However, when it is required to interconnect equipment from various suppliers the problems become glaringly obvious. Some vendors have offered solutions to specific interconnection problems but these generally have to be tackled on a one-off basis leading to complex and possibly expensive solutions, which do not always fulfil the users' requirements. An example of the typical interconnection scenarios that can arise with today's incompatible hardware is shown in Figure 3.2.

The problem outlined so far indicates that there has been a requirement, for some time, to define a standard interface between all computer-based communication systems. If this were to be accomplished it would allow the user to achieve the interconnection of systems regard-

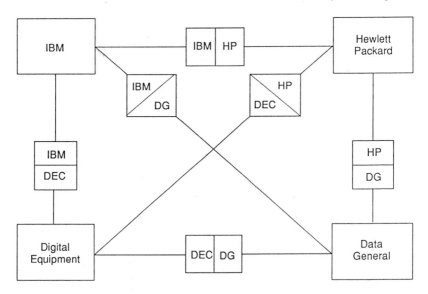

**Figure 3.2   Typical Interconnection Scenario Between Today's
Incompatible Hardware**

less of the source of the individual products. In this way, users would be able to achieve the vendor independence which they all want.

With these problems in mind, ISO decided in 1977 that they would develop a framework for the interconnection of systems. The generic title for this work is Open Systems Interconnection (OSI); and its main principle is that all computers, regardless of make and model, will be able to 'talk to' each other. Both ISO and the CCITT have developed a series of protocol standards with which open systems (systems designed to these standards) will communicate. These protocols are based around a seven-layered architectural model.

The ultimate goal for open systems users is to be able to choose their equipment on the basis of its merit/cost ratio for a particular application without being restricted to systems from a particular supplier. Difficulties with the interconnection of equipment sourced from different suppliers should not exist as they will all be designed to the same standards. Hence, solutions to specific interconnection problems will be redundant which implies savings both in terms of time, effort and

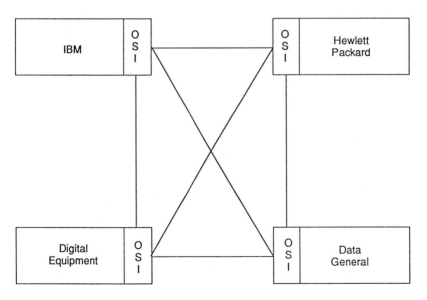

Figure 3.3   Interconnection Utilising OSI Protocols

cost. In Figure 3.3 the interconnection scenario of Figure 3.2 is repeated but, with the use of OSI protocols for the connections between the systems, the simplification is instantly apparent.

However the benefits of open systems to the user will only become available when the equipment vendors adopt the standards. The question that has to be answered is: what incentive is there for implementing the standards?

Once the initial traumatic changeover period — from proprietary to OSI standards — has been overcome, system vendors will start to enjoy some of the benefits of OSI. Because there are no longer any single-vendor domains, each company will be able to sell their products to a far wider market than before. This will lead to competition between the manufacturers to produce the best products at the most attractive prices. Because the OSI architecture will already exist there is no wasted effort expended in the development of new product architectures. In line with this, vendors will be subject to fewer development risks because new equipment will be developed to internationally recognised standards.

In summary, the advantages of open systems *to the user* are that:

— all products conform to the same standards (hence there is greater choice);

— there are no single-vendor domains;

— product purchases are based on their merit for the application rather than whether they interface to existing equipment;

— existing equipment investments are protected because open systems are designed to build in a modular manner;

— there is a reduction in procurement costs because system compatibility is no longer a problem;

— there is no requirement for expensive and complex gateways for systems interconnection;

— improvements in communication will lead to increased distribution of, and access to, information.

The advantages of open systems *to the vendor* are that:

— there are wider markets due to the lack of single-vendor domains;

— there is greater competition to produce the best products at the most attractive prices;

— no effort is wasted in developing new product architectures;

— there are fewer development risks;

— there is wider choice of experienced development staff.

CONCEPTS OF THE OSI REFERENCE MODEL

ISO's first priority was to devise a standard architecture (called the Reference Model for Open Systems Interconnection, currently defined in ISO IS 7498) which could be used to provide a basis for the development of the standards needed, whilst allowing existing standards to be accommodated where relevant. It also provided a method of dividing the work up into manageable portions so that separate teams could work on specific protocols and services.

In practice the Reference Model has been used by some computer manufacturers to provide the framework for their proprietary network architectures. The ISO Reference Model naturally reached a fairly stable state before any standard high-level protocols or services were defined. Thus, computer manufacturers decided that it was sensible to use it and to place their own proprietary protocols within its framework, with a view to replacing these where necessary by the standard versions when they are agreed.

Since layering had proved to be a successful technique for computing and communications functions, the ISO Reference Model was designed as a layered structure consisting of seven functional layers and a physical transmission medium (Figure 3.4). Each layer is self-contained and only the interfaces to adjacent layers, and the service it provides, need to be fixed. Each layer uses services provided by lower layers in order to carry out its functions. Functions within a layer, but in separate systems, communicate with each other using protocols appropriate to that layer. No protocol acts across layer boundaries.

For each layer within the Reference Model, there are two types of standards which are defined in order to specify its operation. These standards address two specific aspects of the service provided:

— the functionality within a specific layer, ie the service provided to the layer above;

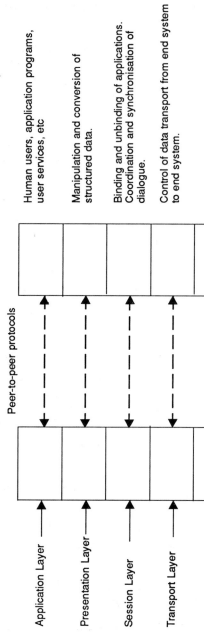

Main features of layers

Human users, application programs, user services, etc

Manipulation and conversion of structured data.

Binding and unbinding of applications. Coordination and synchronisation of dialogue.

Control of data transport from end system to end system.

Routeing and switching.

Reduce errors introduced by physical media.

Provide means to control physical circuits.

Peer-to-peer protocols

physical media for interconnection

Application Layer

Presentation Layer

Session Layer

Transport Layer

Network Layer

Data Link Layer

Physical Layer

Figure 3.4   ISO Seven Layer Reference Model

— the protocols which are utilised within a layer and between peer layers.

At the lower layers, many of these standards are complete, such that many vendors have versions of these standards contained within their products. At the higher layers, work is progressing to the extent where implementations of groups of standards are now beginning to appear.

Within OSI operation there are two basic modes which are supported:

— connection mode — traditional computer communications mode with sign-on, information transfer and sign-off periods during a session;

— connectionless mode — with this mode there is no connection established at the time of transmission. Data is passed to the network with sufficient information to enable a message to be routed to the appropriate destination.

Many of the functions contained within the layers will be applicable to both of these modes of operation while some will only support either of the modes.

PRINCIPLES OF THE LAYERED ARCHITECTURE

The functions within layers are collected into groups which are called entities. These functional groupings can refer to different modes of operation, ie connection and connectionless. Communication between peer entities, both in the same system and in separate systems, is via one or more protocols. The choice of which protocol to use within a layer will be decided by the layer itself, influenced by factors such as:

a) the quality of service required by the layer above;

b) protocol negotiation between peer layers in communicating systems;

c) in the case of connectionless mode the previously agreed choice of protocol (ie the most suitable choice) between peer layers in the communicating systems.

(The following discussion employs the example of an arbitrary layer, N, within a communicating layer stack — see Figure 3.5.)

Entities within a layer-(N) provide a service to the entities in the layer above, ie N+1. In some cases an (N)-entity may require the cooperation

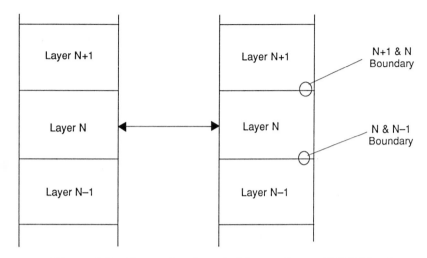

Figure 3.5   Example of an Arbitrary Layer N Within the Communications Model

of another (N)-entity, in order to provide a service to an (N+1)-entity. In this situation entities will utilise an (N)-protocol and an (N–1)-connection. The information exchanged between these (N) peer entities is separated into two parts:

a)   (N)-service-data-unit (N-SDU) — this is the data which (N)-entities require to carry out the functions requested by the (N+1)-entity;

b)   (N)-protocol-control-information (N-PCI) — this protocol information is used by the (N)-entities to coordinate their communication.

These two parts of the data are combined into an (N)-protocol-data-unit (N-PDU).

(N)-SDU + (N)-PCI = (N)-PDU

The data within the PDU may also be separated into smaller pieces, termed segmentation, or combined into larger units, called blocking. The choice of which of these methods to use will be dependent on how to make the most efficient use of the available lower-layer service.

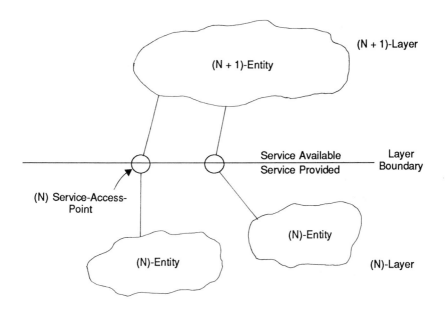

Figure 3.6   SAPs at a Layer Boundary

When an (N+1)-entity requests a service from an (N)-entity it must utilise the appropriate service-access-point (SAP). SAPs are provided on the layer boundary and in the case considered here (N)-SAPs (ie those yielding access to N entities) are located at the N and N+1 boundary (see Figure 3.6). In order to access a specific SAP the (N+1)-entity must utilise the appropriate (N)-address or (N)-service-access-point address, to use its full title.

LAYERS IN THE ISO OSI SEVEN-LAYER MODEL

The physical medium in Figure 3.4 connects the systems involved. It is usually the PTT-supplied network, although it could be a private network such as a local area system. Using CCITT terminology, the boundary is usually the boundary between the data circuit-terminating equipment (DCE) — the modem, for example — and the data terminal equipment (DTE) — ie the subscriber's terminal or computer.

The functions of each of the individual layers in the OSI model are described below.

Physical Layer

The Physical Layer contains functions required for the translation of information or data stream into the form that will achieve transmission, eg electrical impulses, modulation of a carrier wave, etc. This layer provides the electrical and mechanical means for carrying out communications. It is also concerned with procedural aspects of establishing a connection. The service provided by the Physical Layer to the layer above is the most basic, ie a serial data stream.

Many of the standards in this area predate the OSI model and have generally been developed by the CCITT. At this level the standards refer to both modem modulation (eg V.21 and V.27 modem standards), and specifications for interfaces (eg X.21, X.21bis, V.24 etc). Currently, work is under way on the development of ISDN (Integrated Services Digital Network) interface standards (CCITT I-series of recommendations) because the emergence of these networks is likely to have some impact on the future of Wide Area Networks (WANs).

(Note: V.24 lists the handshakes between Data Terminal Equipment-DTEs; X.21 defines the interface for accessing public data networks; X.21bis defines the interface between public data networks and V-series modems.)

Data Link Layer

The principal function of the Data Link Layer is to ensure that data passed to the transmission medium achieves a reliable transfer across the network and hence arrives in an uncorrupted form. Because of the multiplicity of types of linking networks that can be utilised during a transmission it is essential that this function is achieved regardless of the medium being used. The reference model identifies numerous functions for this layer, but two of them in particular are especially important:

— **Framing** This identifies individual messages and groups within a data stream;

— **Error Detection** Physical channels are generally unreliable hence a means of identifying errors is essential.

The recognition that physical transmission mediums are subject to errors, caused by noise and transients produced through inductive means or otherwise, implies that error detection and hence recovery is required. Data link protocols were introduced some time ago to enable errors to be detected and corrected, usually by requesting retransmission. High-level data link control (HDLC) is the most popular example which uses a cyclic redundancy check (CRC) to identify error conditions. CRC employs a mathematical function to generate an error detection code to be passed along with the data stream. Its advantages lie in its extremely high percentage error detection rate and the fact that it introduces relatively little redundancy (bits not conveying information) to the data stream and hence is efficient.

Network Layer

The function of this layer is to create, maintain and eventually terminate a logical path through a network which can be utilised by the Transport Layer. In particular it performs addressing, switching, routeing and facility-selection functions associated with establishing and operating a connection between systems. The intention is to provide a service to the Transport Layer independent of the subnetwork(s) which have been used to transmit the data.

Included in the functions undertaken by this layer are:

— **Routeing**   This decides how to transmit a frame between source and destination using network addresses;

— **Relaying**   This enables data transfer across intermediate subnetworks (see Figures 3.7 and 3.8);

— **Flow control**   This has the ability to match traffic flow with the capacity of the transmission path;

— **Sequencing**   This offers ordered delivery of data across the network connection.

The Network Layer is relatively complex: it has its own internal structure, comprising a number of sublayers. Many of the complications are introduced by the differing requirements of connection-oriented and connectionless services. There are a number of draft proposals in existence from ISO in this area.

The CCITT recommendation X.25 is an interface standard for connection to the public packet switched data networks. This standard covers layers one to three in the OSI model, with the Packet Level (Layer 3) being an example of a possible Network Layer standard. One

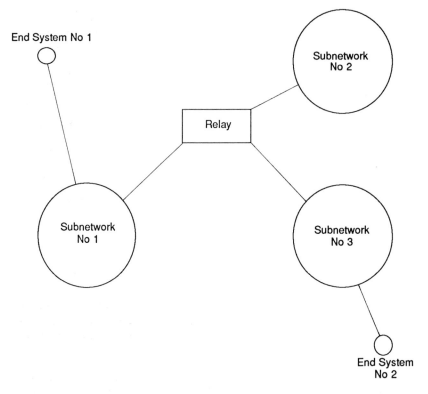

Figure 3.7   The Requirement for Relaying and Routeing

of the basic aims of the Packet Level is to multiplex a number of logical connections over one physical link. At the Physical Layer X.25 will make use of X.21, which defines an interface to operate synchronous mode communications over a Public Data Network (PDN). At its second level, the Data Link Layer, X.25 uses a subset of HDLC.

In the same way that the X.25 recommendation covers the first three layers of the model, the work on ISDN interface (CCITT I-series) and the Local Area Network (LAN) standards (IEEE 802-series and ISO 8802-series) will also cover these layers.

Transport Layer

The main function of this layer is to create a link for the transport of data between the communicating end systems. It accomplishes this by using

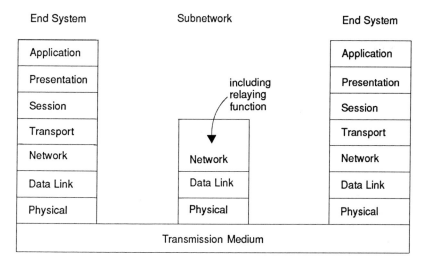

Figure 3.8   Logical View of an Intermediate Subnetwork

the physical transport mechanisms made available by the network. The Transport Layer is the lowest which exists in end systems only (refer to Figure 3.8), and therefore performs functions which are of a purely end-to-end nature.

The Transport Layer provides, in association with the layers below it, a universal data transport service which is independent of the physical medium actually in use. Layers above this service will request a particular class and quality of service, and the Transport Layer is responsible for optimising the available resources to provide the service requested. The quality of a service is concerned with data transfer rate, residual errors allowed and associated items, whereas the class of service covers the different types of traffic which diverse applications require (eg batch or transaction processing).

The services provided fall into one of five categories numbered 0 to 4. The facilities provided by the lower class numbers are in fact subsets of those provided by the higher classes, with class 0 merely being the service provided by the Network Layer, as this adds no functionality to the lower layer. In contrast to Class 0, Class 4 employs a powerful protocol in order to ensure that if any loss, re-order or duplication of data occurs it can be dealt with so that the application will be provided with uncorrupted data. Briefly the functions provided by each class level are:

Class 0 — basic network service;

Class 1 — Class 0 + error detection and recovery;

Class 2 — Class 0 + multiplexing;

Class 3 — Class 1 + Class 2, ie error detection, recovery and multiplexing;

Class 4 — detection of errors not signalled by the lower layers, correction, recovery and multiplexing.

Session Layer

This layer is responsible for establishing and maintaining a relationship between the end users wishing to exchange information. Two applications must have a formal liaison for this purpose, which is called a session. The Session Layer establishes, breaks and maintains this liaison, and ensures that data reaching a system is routed to the correct application. It also ensures that the information exchanged is correctly synchronised and delimited so that, for example, two applications do not try to transmit simultaneously.

The Session Layer is the lowest, in a communicating stack, which deals explicitly with the communication process between open systems rather than just the conceptual interconnection of such systems. The service which it provides to the Presentation Layer is the management of the dialogue between communicating end systems. The functions performed at this layer depend largely on the specific requirements of the application which is in operation. From the range of services which can be provided the following are significant:

— establishment and close down of the connection;

— synchronisation of the data communication process to allow for error checking and recovery;

— negotiation of the manner of interaction between Presentation entities, which can be simultaneous both-way transmission or two-way alternate direction transmission (ie full- and half-duplex respectively).

Presentation Layer

The applications involved in data communication, in an OSI environment, should not be aware of the characteristics which are not relevant to the actual data being exchanged. If this can be achieved it allows the language of the communication to be transparent to the applications which are involved. The Presentation Layer performs the two-way function of taking information from applications and converting it into a form suitable for common understanding (ie not machine-dependent), and also presenting data received to the applications in a form they can understand. The layer provides services which give independence from internal character formats (ie transparency), and consequently it provides machine independence.

When a communication link is established, the end systems will negotiate the encoding rules which are to be utilised during the information exchange. At this stage any conversions which need to be undertaken will be agreed and carried out at this layer.

Two technically aligned standards have evolved in this area: first CCITT recommendation X.409, and then ISO adopted the words and principles for their two-part standard, Abstract Syntax Notation.1 (ASN.1). They utilise the concept of dual levels of syntax, abstract and transfer. At the abstract level data or information is described in terms of its logical structure and its type. When the abstract syntax has been produced it can then be encoded into a data stream which is the transfer syntax. The relationship which the Presentation Layer establishes between the abstract syntax and the transfer syntax is called a Presentation context. Within a connection there can be multiple Presentation contexts which will account for the situation when a number of different abstract and transfer syntaxes are developed and used together.

Application Layer

The Application Layer differs from all the rest in that it is the highest layer of the model and hence does not provide a service to a layer above. It is the source of all data which is to be transported in the OSI environment and ultimately its destination. All the other layers in the model exist to support this layer which in turn exists to provide the information exchange functions between user application processes. It is here that a decision is made whether to treat the communication as a file transfer, virtual terminal session or an MHS interpersonal message.

Because of the requirement for multiple specific applications to reside at this layer a formal structure was required. This would specify exactly how application associations (ie communications between applications in separate end systems) should be formed using particular elements. Unfortunately this area has been the source of some confusion mainly because the original modelling concepts which were proposed have since been changed in order to encompass new ideas.

Initial ideas surrounded the use of two sets of basic elements with which associations could be formed. The first set of elements relate to the establishment and release of connections forming a basic 'tool kit' called Common Application Service Elements (CASEs). The second set are specific to certain jobs or activities and are called Specific Application Service Elements (SASEs).

CASE functions, as presently defined, are further subdivided into two categories:

1 — Association Control
These services allow an application to 'Begin', 'End' and 'Halt' Presentation connections.

2 — Commitment, Concurrency and Recovery (CCR)
These services allow an application to have control over the 'atomic actions' of a connection.

SASE functions cover specific jobs, tasks or activities. So far, ISO work in this area has been on bulk file transfer (known as FTAM), job execution upon remote systems (JTM), and the use of a terminal type which is application-independent (VT). It is likely that other ISO-developed SASEs will be developed to cover other tasks and it is also likely that the private sector will develop their own SASEs for industry-specific applications.

Message Handling Systems based on the 1984 X.400 recommendations do not fit into the CASE-SASE applications model. Instead the concepts of the Remote Operations and Reliable Transfer Service (RTS) are employed to give the application entities, SDEs and MTAs, a simplified interface to the Session Layer with a minimal Presentation Layer service.

This original model has now been changed and is re-specified as the generic Application Layer structure within a new standard, ISO/DIS 9545. In the OSI environment, communication between application processes is now represented in terms of an association between Application Entities (AEs) residing within the Application Layer using a Presentation Layer connection. A user's local process will interact with the Application Layer via a User Element (UE) which has an AE associated with it (see Figure 3.9). The interaction between AEs is described in their use of Application Service Elements (ASEs) which are subdivisions of the total functionality of the whole AE. An ASE is a grouping of functionality which will support a typical application. Each ASE represents a different kind of work which the user requires to be performed. ASEs come in basically two types:

> Re-usable ASEs — these are analogous to the concept of CASE functional elements, as specified in the early model, because they provide generic services which are 're-usable' in many types of association. Examples of these would be ACSE, CCR, ROSE and RTSE.

> Application Specific ASEs — these are analogous to the SASE functional elements, as specified in the early model, because they are specific to a particular job or function. The application-specific ASEs may have their functionality divided still further into other ASEs which deal with specific aspects of a service to be provided. Examples of these would be FTAM, JTM, VT and MHS. The MHS ASE, from the 1988 recommendations, has been subdivided into other ASEs, such as:

> — ASEs used for access to the Message Transfer (MT) service;

> — Message Submission Service Element (MSSE);

> — Message Delivery Service Element (MDSE);

> — Message Administration Service Element (MASE).

The combination of ASEs used in a specific communication scenario is termed an application-context. The choice as to which application-context is to be used will be dependent on the type of work that the application process has requested to be completed via the UE.

Although the situation with two successive models for the Application Layer structure appears unnecessarily complex it is not a source of potential worry. It should be emphasised that a change in the modelling concepts does not definitely imply changes in real implementations. The model is only a means of conveying the concepts behind a more complex operation which in this case remains unchanged although the model has been updated.

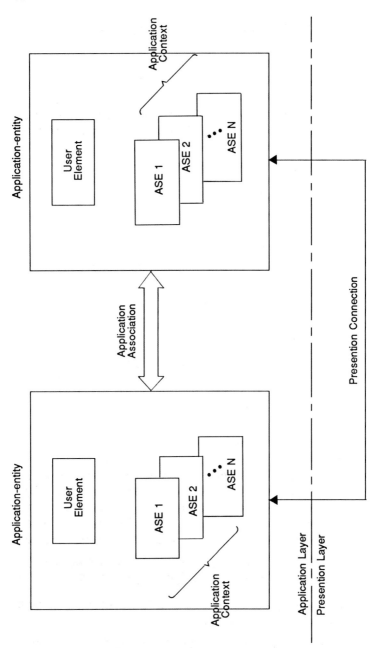

Figure 3.9   The Application Layer Model

The following sections discuss the ISO developed ASEs now available.

File Transfer Access and Management (FTAM)

FTAM is the ISO OSI standard designed to allow open systems within a distributed computer system to transfer, access and manage files. Some examples of the types of 'real time' applications which might employ this standard are:

— complete file or bulk file transmission;

— distributed database systems;

— file servers and remote clients;

— remote management of filestores.

The FTAM standard is structured into three parts which define:

— the Virtual Filestore;

— the File Service;

— the File Protocol.

The *Virtual Filestore* is concerned with three specific areas:

— the Filestore Model;

— the Actions of the Filestore;

— the set of Attributes of files.

The Filestore Model defines a descriptive view of real-file stores, which is independent of the method of implementation. The structure of a file in abstract terms is outlined, using such file attributes as a unique filename and subordinate to this a logical 'tree structure' of data units each of which can be identified. This Virtual Filestore also forms a general plan onto which real filestore implementations can be mapped.

The Actions of the Filestore are those which can be performed on the file, its attributes or data units. These are split into two distinct sections:

— actions on complete files (create, select, change/read attribute, open, close, delete, deselect);

— actions on the content, ie its units of data, of a file (locate, read, insert, replace, extend, erase).

The *File Service* defines the set of service primitives and associated details which refer to actions performed upon the Filestore Model. The exchange of these primitives occurs during the association of two application entities.

The elements of the File Service are split into regimes which refer to specific groups of actions being performed. Three such regimes are defined within the File Service:

— File Transfer Access and Management (FTAM);

— File Open;

— Data Transfer.

The *File Protocol* defines the actual interaction taking place between application entities in order to use the elements of file service.

Job Transfer and Manipulation (JTM)

One use of an open system environment will undoubtedly be to process jobs entered from a remote location. JTM is historically derived from the concepts of Remote Job Entry (RJE) applications, but is much more comprehensive and powerful in its approach. The standard was designed to allow one entity, the Initiator, to produce a Work Specification which can then be sent for Execution by another open system. The Work Specification specifies the operations to be performed and the sources of the data, which may be local or resident on another Source system.

Job entry is only one aspect of remote job processing. Others include:

— user enquiry on the status of jobs being processed;

— operator manipulation and enquiry;

— cancellation of jobs;

— operator-to-operator communications;

— system message transfer;

— user file transfer.

Virtual Terminal (VT)

In an open systems environment it will be possible for a wide range of terminals to be interconnected with an equally wide range of computer systems. In a closed, single-vendor, domain it is possible to ensure that a standard terminal or terminal handling method is employed. In an open systems environment this is not possible and a standard means of achieving terminal interaction is required so that application independence from terminal characteristics can be accomplished.

The idea of a Virtual Terminal (VT) has been developed to meet these requirements. The VT standard is structured into both service (VT Service = VTS) and protocol (VT Protocol = VTP) parts. It provides a means for VT users to communicate, regardless of their local terminal type, using a common VT protocol. A VT user can be a terminal or a process, with terminal-to-process, terminal-to-terminal, and process-to-process working all allowed. The terminal-to-process interaction, mainly for data entry, would seem to be the major application for VT standards. The latter two interactions will probably be accomplished by other OSI applications: X.400 Interpersonal Messaging (IPM) for messaging between terminals; FTAM for terminal-to-terminal bulk transfer; and Open Distributed Processing (ODP) for process-to-process working.

VTS attempts to overcome the compatibility problems of differing terminal types by using a shared conceptual communication area (CCA) which contains a wide ranging set of generic terminal functions (such as clear screen, background colour, foreground colour). The functions of each user's terminal will then be mapped onto this range of generic functions; in this way, commonality between the functions of each terminal is achieved.

VT Service Classes

In order that the VTS is able to support terminal functions required by applications, both now and in the future, a number of classes of service will be catered for. These service classes are designed to meet the requirements of a range of specific applications. Those defined so far by ISO are the Basic Class and the Forms Class, although other classes may be provided for to cope with image, graphics and mixed mode (ie a combination of other classes):

Basic Class VT Service — This service will allow the functions achieved by simple PCs to be mapped, eg colour modes, full-character repertoire and simple text editing features.

Forms Class VT Service — Data entry to application processes is

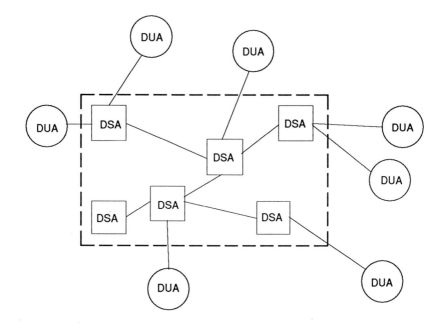

Figure 3.10   The Directory Service Functional Model

a common activity within administration departments and this class was devised with this in mind. It allows local applications to map the data entry on their screen format to that of the application which is receiving and processing the data.

OSI Management

In order for any network to operate satisfactorily it has to be managed effectively. This is particularly true of an OSI environment because the range of communication and application options is so extensive. Effective management will ensure that resources are allocated in the most appropriate way, fault conditions can be recovered from, accurate account information is collated, and security measures are applied where applicable.

The management framework being proposed for OSI systems will provide the necessary sensors, at all levels, required to gather the

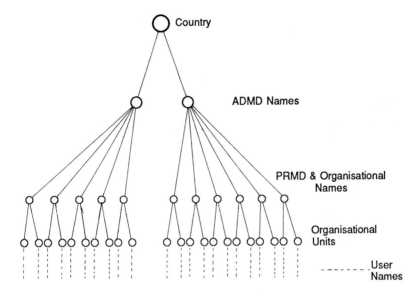

Figure 3.11   An Example of a Directory Information Tree

information to be able to manage such a distributed environment. The framework specifies three forms of management information exchange:

Systems Management This is considered as the normal method of managing resources within an Open System. Systems Management communication is provided via Application Layer protocols which are in turn provided by the elements of the Management Information Service (MIS). These protocols allow information to be exchanged which relates to the monitoring, control and coordination of resources within an Open System.

(N)-Layer Management In certain cases it is required to monitor the operation of a specific layer, ie management of an (N)-layer. This mechanism is provided to account for this situation and allows for the monitoring, control and coordination of a specific layer.

(N)-Layer Operation Certain aspects of management will already be performed within the normal operation of layers and these will refer to a specific instance of communication. Hence a mechanism is required to monitor this information.

The Management Information Service (MIS), mentioned earlier, is the generic title given to the range of Specific Management Information Services (SMISs), based at the Application Layer. SMISs are subdivisions of the overall management activity separated into functional groups. At present there are five such services defined:

— Fault Management;

— Configuration Management;

— Security Management;

— Accounting Management;

— Performance Management.

The SMISs make use of a common tool-kit of services which are provided by the Common Management Information Service (CMIS). The CMIS provides the range of primitives required by the SMISs to manage the resources of an Open System. The CMIS in turn makes use of two other Application Service Elements (ASEs): Association Control Service Elements (ACSEs); and the Remote Operations Service Elements (ROSE).

OSI Directory Service

The OSI Directory Service is a specialised distributed database facility which will hold all of the addressing information required to support communications in an OSI environment. Because such a service is crucial to the successful operation of worldwide Open Systems Interconnection, both CCITT and ISO quickly realised the need for a coordinated approach to its definition (CCITT X.500 Series & ISO 9594/1-8). The CCITT work was born out of a need to have a globally interlinked directory for Message Handling Services, whereas ISO's need stemmed from work on OSI management.

The scope of the work on the Directory Service covers both the information held within it and the means with which a user will access this, ie access protocols.

Figure 3.10 indicates the functional model for the Directory Service. The Directory Information Base (DIB) is the collection of all the directory information; this will, in practice, be distributed on a worldwide basis. The DIB will be distributed amongst the Directory Service Agents (DSAs) which communicate with each other via a protocol known as Directory System Protocol (DSP). The user of the directory service is represented by a Directory User Agent (DUA) which can communicate with DSAs via the Directory Access Protocol (DAP). Both DSP and DAP are based on the use of X.400 1988 Remote Operations Service Elements or ROSE. For organisational purposes the DIB will be split into management domains which will be under the control of the group which maintains it. Two such types of domain are defined:

— Administration Directory Management Domains (ADMDs)- provided by either a service provider or PTT;

— Private Directory Management Domains (PDMDs)- would be provided by a private organisation, such as a large company.

The Information in the DIB is not in the form of a flat file but rather uses the concept of an hierarchical tree structure which is called the Directory Information Tree (DIT). Figure 3.10 shows an example of a DIT representing the information required for X.400 addressing.

OSI Security

This will be produced as an addendum to the original reference model document and sets-out to identify the security functions which should be associated with the layers of the OSI model. These functions can be carried out at Layers 3, 4 or 7 of the model and work is currently underway by the relevant groups to include these concepts in the base documents for these layers. The 1988 CCITT X.400 drafts and new ISO MOTIS documents (formed jointly with the CCITT) now carry sections on security and how it should be incorporated in a Message Handling System (MHS).

CONCLUSION

This chapter has provided a flavour of the general concepts and reasoning behind Open Systems Interconnection (OSI). The following chapter provides an in-depth discussion of the technical concepts of the X.400 MHSs which are based on the principles of OSI.

4 X.400 — A Technical Introduction

INTRODUCTION

In Chapter 3, the principles of the Open Systems Interconnection (OSI) seven-layer model were considered. The uppermost layer of this model is called the Application Layer; the processes which reside there provide the means of information transfer, between user applications, in the OSI environment. Message Handling Systems (MHSs), based on the CCITT's X.400 recommendations, are an OSI application which provide store-and-forward message transfer facilities to the user.

The X.400 series of recommendations comprises eight closely related documents which define a generic architecture for medium independent computer-based messaging systems. As with all international standards and recommendations, X.400 is riddled with jargon which will confuse the casual observer. It is the intention in this chapter to break down this barrier and to introduce the basic technical concepts of these recommendations in a clear and concise manner. This will be by no means an exhaustive study of the topic; rather it will provide a sound knowledge base upon which further research can be carried out where required.

DEVELOPMENTS IN STANDARDISED MESSAGING

The development of computer-based Message Handling Systems (MHSs) began in the mid-1960s with a number of groups developing systems to allow the exchange of electronic messages within computer networks. These developments gave rise to product manufacturers producing their own mail system application packages for existing computer hardware. This was a perfectly acceptable solution, providing the systems to be interconnected were produced by the same manufacturer, ie there was a single vendor.

The development of standardised solutions for electronic messaging systems began in the early 1980s with the realisation that the true benefits of electronic mail are inaccessible without international standards, ie full interconnection and integration. The International Federation for Information Processing (IFIP) developed a basic model of a store and forward message transfer system. It was these concepts which were rapidly adopted and developed by three major standards groups, ECMA, CCITT and ISO. The development work of these respective organisations differed at detail levels because of the service spectra which they were expected to address, eg CCITT, administrations and service providers, ISO, the private messaging environment.

The CCITT have a traditional responsibility for the standardisation issues surrounding international telecommunications. Their work is carried out in four-year study periods at the end of which recommendations are published in the form of books. The colour of these publications indicates the study period from which the recommendation emanates; for example, the X.400 series was produced during the 1981–1984 period and is contained in a book with a red cover.

When in 1984, the CCITT ratified their set of eight MHS recommendations (numbered non-sequentially X.400 to X.430) it was decided that the ISO Message Oriented Text Interchange System (MOTIS — ISO's version of the message handling work) should adopt the same words and principles wherever possible. MOTIS, however, could be considered as a compatible functionally extended version of X.400, extending it into the private messaging domain. This is specifically indicated by the ISO inclusion of links between private messaging domains as CCITT do not specify this particular link.

After ratification of the X.400 recommendations it soon became obvious that they contained a number of errors and ambiguities which needed correction in order to produce a usable series of documents. To accomplish this the CCITT embarked on the production of a series of *Implementors' Guides* which attempted to highlight these problems.

Ultimately, however, it was the intention of the CCITT to have a series of base documents which were as near perfect as possible. To this end, development on the messaging topic continued during the subsequent study period (1985/88) and has culminated in the production of a new series of documents. These are considered in Chapter 6 which looks at the future of message handling.

X.400 RECOMMENDATIONS

The CCITT's X.400 recommendations are a series of eight closely related documents:

X.400 — specifies the network model comprising:

— user agents;

— message transfer agents;

— message transfer;

— interpersonal messaging services;

— details of the protocol layering of an MHS.

X.401 — describes the basic service elements and optional user facilities of the MHS. This recommendation gives an overview of the Interpersonal Messaging (IPM) and Message Transfer (MT) services which provide the above functions.

X.408 — specifies the rules for converting one information type to another, eg telex character codes to IA5 (ASCII). This recommendation was largely left for further study in 1984; however, this has been remedied in the 1988 X.400 recommendations.

X.409 — defines the presentation transfer syntax rules used by Application Layer protocols in an MHS. Within the Open Systems Interconnection (OSI) architecture this defines a Presentation Layer (Layer 6 of the model) syntax which is used to represent information exchanged between applications.

X.410 — details the facilities provided by Remote Operations and the Reliable Transfer Server which make up the lowest layers of the sub-layered MHS architecture.

X.411 — specifies both the service and protocol for the Message Transfer Layer (MTL) within the MHS model.

X.420 — specifies the content protocol for interpersonal messaging services and includes the format for memo headers and multi-part body types.

X.430 — details the specialised means of access for teletex terminals which allows them access to the functions of the Interpersonal Message (IPM) and Message Transfer (MT) services.

For a newcomer to the subject the X.400 recommendation serves as a suitable introductory document to message handling topics. It defines the main components of a system and their basic interaction without descending too deeply into the complex subject of protocols.

WHAT IS A MESSAGING HANDLING SYSTEM?

An X.400 Message Handling System (MHS) may be viewed as a store-and-forward electronic mail system. Note the term store-and-forward: these are not real-time systems and hence are not suitable for real-time (ie immediate response) applications. X.400 systems allow people or computer application processes to communicate by submitting and receiving messages which can consist of any structured or unstructured data. Because the X.400 Message Handling System (MHS) recommendations have been internationally agreed they can provide a standardised electronic mail system for the integration and interconnection of all today's disparate office systems and equipment.

BASIC BUILDING BLOCKS

It was stated earlier that X.400, in common with all International Standards and recommendations, is full of jargon which requires some definition before it is possible to grasp some of the basic concepts. Consider the individual components of an X.400 MHS (as indicated in Figure 4.1):

MESSAGE TRANSFER AGENTS (MTAs)

These are store-and-forward message exchanges whose functions are analogous to those of sorting offices within the public postal service. They are interconnected and are known collectively as the Message Transfer System (MTS).

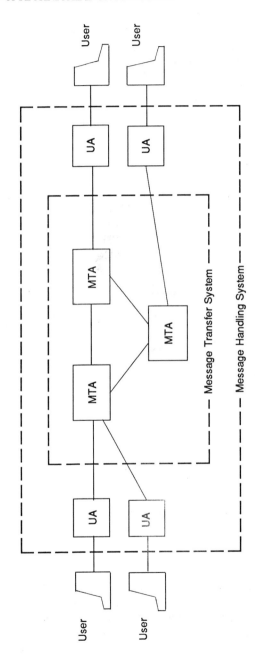

Figure 4.1   The Basic Components of an MHS

The interconnection of MTAs is via any media capable of supporting full OSI network protocols.

USER AGENTS (UAs)

These provide the user interface to the MTS and may include local editing, temporary storage and archive facilities. The UA accepts a message from the user and then employs the MTS to route the message to the recipient.

USER

This, in X.400 terms, is not a human user; rather it is the equipment or service which is using the MHS to achieve message transfer. It may be a dumb terminal, local host or a third-party electronic mail facility connected through a public data network.

The complete interconnection of UAs and MTAs is called the Message Handling System (MHS).

It must be emphasised that the X.400 recommendations do not specify the manner in which the components of an MHS are to be implemented. The MTAs and UAs are merely groups of functionality placed together because they appertain to a particular type of function, ie the UA deals with user-related functions while the MTA is concerned mainly with the message transfer function.

THE MESSAGE HANDLING SYSTEM MODEL

The store-and-forward message handling model is analogous to the familiar postal service which employs a form of sorting office and delivery agents. With this model, letters or messages are placed into the system and their delivery is determined by a number of factors. The address on the message envelope will dictate the destination to which a letter is delivered while the time taken for this process to be accomplished will depend on: a) grade of delivery (1st or 2nd class); and b) how busy the service is (seasonal variations, such as Christmas).

In an MHS the destination of the message is also determined by its address placed upon the delivery envelope; though in this case it is an electronic protocol envelope rather than a physical realisation. The grade of delivery of an X.400 message can also be specified between bounded limits, ie delivery within 40 mins, two hours, etc. Ultimately though there is no fixed relationship between the time when the message is sent and the time when it arrives, outside the limits mentioned earlier. The traffic being handled by the system at any particular time will dictate the priority of messages in addition to the grade of delivery specified. The basic MHS model is shown below, in Figure 4.2, and from this it can be seen that the process of message delivery splits conveniently into three distinct phases.

The three phases in the transmission of an X.400 message are dealt with below.

Message Submission

The user through interaction with the 'originating UA' will produce a message for despatch. This part of the operation is analogous to a person producing a letter and depositing it in their post box ready for collection. The UA will then 'submit' the message to its MTA in an envelope which contains all the information required by the MTS to route the message to its intended recipient.

Message Relay

The MTA will analyse the information on the envelope in order to determine the route across the MTS. MTAs can be considered as being analogous to sorting offices in the postal service because they look at the envelope information in order to determine the intended recipient. The

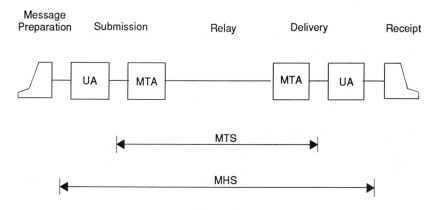

Figure 4.2   The Message Handling System Model

actual process of message relay may involve one or more MTAs and each in the chain will time-and-date stamp the message envelope so that a record of its route across the MTS will be maintained. In a similar manner to specifying a class of service for a letter (eg 1st or 2nd class stamp) it will be possible to specify the urgency of delivery in a practical X.400 system, eg ranging between urgent and non-urgent.

Message Delivery

The message will eventually progress through the system until it reaches the MTA associated with the intended recipient. When this is achieved, the MTA 'delivers' the message to the recipient's UA in a similar manner to how the postman delivers a letter to the recipient's letter box. The final stage of the transfer will be complete when the message is received; this is achieved when the user accesses the message in order to read it.

Introducing the MHS model with this simple analogy is a reasonable way of conveying the basic concepts. The terms to remember here are *submission, relay* and *delivery.*

MESSAGE HANDLING SYSTEM SERVICES

In order that X.400 MHSs can achieve application independence, the services provided by them are separated into two distinct groups. In this way the functions of the service which appertain to a specific user application are separated from those which provide the underlying message transfer function. The two types of service provided by the MHS are:

— Message Transfer Service or MT-Service (never shortened to MTS where there may be confusion with the Message Transfer System acronym);

— User Services.

Message Transfer (MT) Service

This is the generic service which is operated between MTAs and UAs to provide an application-independent, store-and-forward message transfer service. This service can carry any form of structured or unstructured data as it is only concerned with the envelope information unless it is expressly requested to carry out some form of content conversion. It is this ability to carry any form of data that allows X.400 to act as a basis for applications other than just straightforward text messaging.

The basic MT service enables UAs to access and be accessed by the MTS in order to exchange messages. When a message is to be transferred it is given a unique message identifier; hence if the message cannot be delivered, the originating UA can be informed.

The MT service consists of a number of service groups which except for the 'basic' group are optional. These groups are in turn subdivided into a series of service elements. Each of these service elements performs a specific function within the MT-service. The service groups are formed from service elements which together perform a specific aspect of the MT-service. The service groups are:

BASIC—these are concerned with the unique identification of the message, date/time of message delivery/submission and the type of contents;

SUBMISSION AND DELIVERY — these are concerned with the specification of the type of delivery which is required (eg speed, number of recipients, etc) and whether any delivery notification is applicable;

CONVERSION — whether or not message content conversion is required (dealt with in more detail later);

QUERY — to establish whether it is possible to deliver a message, ie whether it is in a form suitable for display on the recipient's terminal;

STATUS AND INFORM — this service group allows messages with certain attributes to be only delivered to specific UAs and gives the UA the capability to indicate to the MTS that it is not ready to accept message delivery.

After looking at the general groups within the MT-service the Table 4.1 is a complete list of the features provided by this service (a full explanation of the features is given in recommendation X.400, sections 4.1.1 to 4.1.5.2).

Most of the list of services specified in Table 4.1 will involve the user in making a decision (ie whether or not to have a particular service). The manner in which these are presented to the user along with the elements

Service Group	Service Elements
Basic	Access management Content type indication Converted indication Delivery time stamp indication Message identification Non-delivery notification Original encoded information types indication Registered encoded information types Submission time stamp indication
Submission and Delivery	Alternate recipient allowed Deferred delivery Deferred delivery cancellation Delivery notification Disclosure of other recipients Grade of delivery selection Multi-destination delivery Prevention of non-delivery notification Return of content
Conversion	Conversion prohibition Explicit conversion Implicit conversion
Query	Probe
Status and Inform	Alternate recipient assignment Hold for delivery

Table 4.1

of the User Service is particular to each implementation. For instance, it may be possible to specify default values for many of the services which remain unchanged unless specifically requested to do so. Such defaults may be part of a particular user's system profile. These details, however, are outside the scope of this recommendation as they will have no effect on the externally measurable qualities of each system.

Although the MT-service generally operates purely as a means of transferring messages, there is a special case where it takes a more active role in the formatting of the message in order to achieve delivery. 'Content Conversion' is where the MT-service will re-format the content part of a message so that it can be delivered to a non-X.400 terminal. An example here might be the delivery of an X.400 message to a telex machine or a teletex terminal. Access to these facilities will be provided by the PTT or service provider controlled X.400 services utilising special conversion gateways. The recommendations concerning the conversions to be made between information types (eg IA5 to telex, etc) are incomplete and specified for further study. This is not such a problem however, as the concepts of such conversions are well understood. The types of conversion to be offered by an MHS, however, have already been finalised. These are shown in Figure 4.3.

User Services

While the MT-service is comprehensive and complete there has only been one user service defined so far. This is the Interpersonal Messaging Service (IPM service) which provides its users with electronic mail/ messaging facilities similar to those of existing public electronic mail services. Each user service will be defined to accomplish a specific user task or application (such as electronic messaging) and for each one defined a new class of UA must also be defined.

The design of these new UAs is not hindered by having to be compatible with those of existing user services. They merely have to be able to interface successfully with the underlying message transfer system. Some potential new user services are discussed at the end of this chapter to indicate the underlying flexibility of these systems.

Interpersonal Messaging Service

The Interpersonal Messaging (IPM) service provides its users with electronic mail/messaging facilities. The service is built upon the MT-

From \ To	TLX	IA5 Text	TTX Basic	TTX Optional[1]	G3 Fax Basic	G3 Fax Optional[1]	TIF0 Basic	TIF0 Non-basic[1]	Videotex	Voice	SFD	TIF1 Basic	TIF1 Non-basic[1]
TLX[5]	–	a	a	a	a	a	a	a	b	b	a	a	a
IA5 Text	b	–	b	b	a	a	a	a	b	b	a	b	b
TTX Basic	b	b	–	a	a[4]	a[4]	a[4]	a[4]	a	b	b	a	a
TTX Optional[1]	b	b	b	–	a[4]	a[4]	a[4]	a[4]	a	b	b	a	b[2,3]
G3 Fax Basic	c	c	c	c	–	a[4]	a	a	c[6]	c	c	a	a
G3 Fax Optional[1]	c	c	c	c	b	–	b	b[2,3]	c[6]	c	c	b	b
TIF0 Basic	c	c	c	c	a[4]	a[4]	–	a	c[6]	c	c	a	a
TIF0 Non-basic[1]	c	c	c	c	b	b	b	–	c[6]	c	c	b	b[2,3]
Videotex	b	c	b	b	a[7]	a[7]	a[7]	a[7]	–	FS	FS	FS	FS
Voice	c	c	c	c	c	c	c	c	c	–	c	c	c
SFD	b	b	a	a	a[4]	a[4]	a	a	b	FS	–	a	a
TIF1 Basic	b	b	b	b	a[4]	a[4]	a	a	b	b	b	–	a
TIF1 Non-basic[1]	b	b	b	b[2,3]	a[4]	a[4]	b	b[2,3]	b	b	b	b	–

Figure 4.3   MHS Encoded Information Type Conversions

–   No conversion.
a   Possible without loss of information.
b   Possible with loss of information.
c   Impractical.
FS For further study.
1   Specified in the relevant Recommendations.
2   No information is lost if originating and recipient terminals have the same optional functions.
3   Information may be lost if the originating terminal uses optional functions that the recipient terminal lacks.
4   Information may be lost due to the difference between the printable and reproducible areas.
5   The WHO ARE YOU character is assumed to be a protocol element used for communicating with the telex terminal and not part of the message's content.
6   It may be possible with loss of information, if the recipient terminal has the capability of the photographic type of information.
7   When converting videotex, colour information may be lost.

service and is provided by a special class of UAs called IPM UAs. The service also allows intercommunication with telex and telematic services such as facsimile and teletex via gateways or access units. In addition to these, X.400 links will also be established with other forms of proprietary and private electronic mail systems. This will eventually mean that an IPM user will be able to exchange messages with the whole range of other text communications services.

IPM UAs make use of the basic group of MT facilities while also permitting the optional ones to be requested. In addition to these service elements IPM UAs also provide other capabilities which make up the IPM service. These additional functional groups are:

COOPERATING IPM UA ACTION — These are service elements which involve the IPM UA in some form of action upon the message, eg receipt notification, auto-forward indication;

COOPERATING IPM UA INFORMATION CONVEYING — These service elements are concerned with the identification of the originator/recipient and status of a message.

With the addition of the service groups above the complete range of IPM facilities is shown in Table 4.2.

It is worth emphasising again that in a typical implementation these service elements would be selected by the user filling-in the appropriate fields on the display format. It is probable that only a subset of these service elements would be displayed to the user, with the others being set to default values. This kind of detail is not considered within the X.400 recommendations as it is the concern of the man–machine interface. This level of detail will be particular to each individual implementation.

It can be seen from the list provided that the facilities offered by the IPM service are similar to those offered by existing electronic mail/messaging systems. The major advantages of X.400 systems is that they offer worldwide standardisation and ultimately interconnection for electronic messaging. In addition to the facilities listed above, the IPM UA may also perform local functions such as preparation/editing and filing/retrieval of messages. Such functions are, however, outside the domain of the standard and are considered to be specific to each implementation as they will have no effect on the externally measurable qualities of a communicating system.

Service Group	Service Elements
Basic	Basic MT service elements IP-message identification Typed body
Submission and Delivery and Conversion	As per MT-service table
Cooperating IPM Action	Blind copy recipient indication Non-receipt notification Receipt notification Auto-forwarded indication
Cooperating IPM UA information Conveying	Originator indication Authorising users indication Primary and copy recipients indication Expiry date indication Cross-referencing indication Importance indication Obsoleting indication Sensitivity indication Subject indication Replying IP-message indication Reply request indication Forwarded IP-message indication Body part encryption indication Multi-part body
Query	As per MT-service table
Status and Inform	As per MT-service table

Table 4.2

The IPM UA is basically providing its services by attaching a header to the message content parts (see Figure 4.4). This conveys all the information necessary to allow the IPM service to operate. The information contained here will contain both general information (ie from, to, subject, etc) and information about actions to be performed (eg courtesy copy, recipient lists, etc). A message may consist of more than one body part and each may be of a different content type, eg text, image, voice, etc (again see Figure 4.4). When a message has been prepared for transmission the header and the body parts will be enveloped by the MT service for despatch to the recipient.

CONFIGURATION OF X.400 SYSTEMS

So far the two major functional blocks of an MHS have been introduced:

— Message Transfer Agent (MTA);

— User Agent (UA).

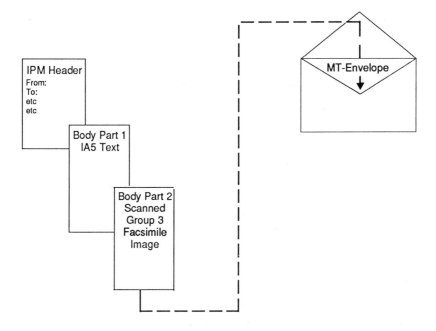

Figure 4.4   The IPM Message Structure

These can be configured in a practical system in a number of ways. Probably the most popular configuration will be for the UA and MTA to reside on the same hardware, ie co-located. In this situation both elements will probably be implemented in separate software packages from the same vendor. With this arrangement the interface between the UA and MTA does not have to be standardised as it has no externally measurable effect on the operation of the system. Hence, a proprietary interface is perfectly acceptable.

A further option with the implementation of X.400 systems is to have the UA located remotely from its MTA, ie implemented on separate pieces of hardware. With this arrangement it is possible, indeed likely, that the two products will be from different manufacturers. In this case there is a clear requirement for a standardised interface between a remote UA and its MTA. Within the X.400 recommendations this interface has been provided in order to allow multi-vendor products to be interconnected.

Figure 4.5 indicates the options which are possible when configuring practical X.400 systems.

REMOTE USER AGENT ACCESS

Provisions are made, within the model, for the option of remote UA operation, with 'remote' meaning that the MTA and UA are not co-located. In such cases the remote or stand-alone UA is split into two functional entities:

— the User Agent Entity (UAE);

— the Submission and Delivery Entity (SDE).

The UAE provides the user with the normal UA facilities and functions, while the SDE provides the UAE with access to the facilities of the Message Transfer (MT) service. Whereas message transfer between MTAs is based on the store-and-forward principle SDE and MTA interaction is interactive in the same way that it would be when the MTA and UA are co-located.

One of the most glaring omissions from the original 1984 X.400 recommendations is the lack of a message store buffer which can be used when a UA is remote from its MTA. When a message is received by the recipient's MTA, it is always delivered directly to the UA or remote UA.

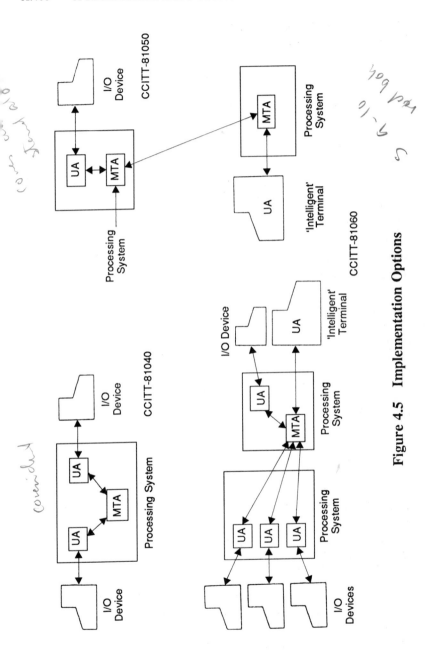

Figure 4.5 Implementation Options

This mode of operation is perfectly acceptable when UA and MTA are co-located because, when one element is active, the other is as well. Hence when the UA receives a message it will store it in its own message store until the user becomes active and accesses it. In the case of the remote UA this instant delivery of a message from the MTA to the UA assumes that it is always attached and active. This is a fair assumption if the remote UA is implemented on a minicomputer, for example, which is active 24 hours-a-day. But if it is implemented on a microcomputer it is not likely that it will remain active 24 hours-a-day or even every day (eg if the human user is on holiday). In this situation the messages will be held at the MTA until such time as the remote UA becomes 'live'. At this point the MTA will attempt to deliver all the messages for the user to his UA.

This situation is still acceptable unless the UA happens to have been off-line for some time. Now the MTA attempts to deliver a large number of messages, hence 'flooding' the storage area. If there is insufficient storage area available there will undoubtedly be a loss of some messages. This serious problem can be overcome if the user has a personal message store buffer resident at the MTA. This would allow the user to selectively receive messages in an orderly manner rather than being 'flooded'.

This problem reduces the potential usefulness of the remote UA approach, a fact which was quickly realised, particularly as the move towards ever more complex microcomputers continues. The solution to this situation is provided by the 1988 version of the X.400 recommendations which include message store concepts (see Chapter 6).

TELETEX ACCESS

Access to systems/services other than teletex from X.400 will be provided via gateway facilities. The functionality of such facilities is not standardised as generally their operation will (or should) be invisible to the users. Teletex was deemed to be a special case in that it is also a standardised service in which the terminal types, largely computer-based, have sufficient functionality to make use of the facilities provided by the Interpersonal Messaging Service (IPMS). Also at the time when the X.400 recommendations were being formulated it was felt that teletex would have a great future and quickly establish a large user base. In practice this has not been the case, except in Germany its country of origin.

Teletex access is defined in recommendation X.430 and uses the concept of an access unit, naturally known as a teletex access unit or TTXAU for short (see Figure 4.6). The purpose of the TTXAU is to aid the user of a teletex terminal in gaining access to the features of the IPMS. Basically the TTXAU is an intelligent gateway facility which appears as an IPM UA to the MT-service while providing an interface to teletex on the other side. The TTXAU may also provide Document Storage (DS) which allows messages received from the Message Transfer System (MTS) to be stored under a user's ID until accessed by the teletex user. The teletex terminal and the TTXAU communicate via a protocol known as P5.

PROTOCOLS WITHIN X.400

It has already been stated that message handling is an application process that resides within the Application Layer of the OSI model. Inside this layer the MHS functional entities are accommodated, ie the UAs, MTAs and SDEs. In order to achieve a logical separation between the services provided and the protocols which they employ, the concept of sub-layers, within the OSI Application Layer, was introduced (see Figure 4.7). The first of these layers is the User Agent Layer (UAL) within which the User Agents (UAs) reside. The second layer is the

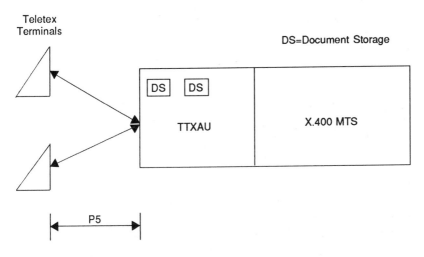

Figure 4.6   TTXAU (Teletex Access Unit)

Message Transfer Layer (MTL) within which the Message Transfer
Agents (MTAs) and Submission and Delivery Entities (SDEs) reside.
The main reasons for the introduction of these sub-layers are:

— to allow a boundary to separate groups of similar functionality;

— to minimise the interactions across layer boundaries;

— to allow different protocols within layers to be used without
   affecting each other.

The use of sub-layering concepts allows various protocols to operate
within the Application Layer without interference. X.400 actually
specifies four MHS protocols, P1, P2, P3 and P5, all of which except for
P5 (teletex access protocol) exist at the Application Layer of the OSI
model. The functions of these protocols are as follows:

P1 — provides the basic message relay envelope function between MTAs;

P2 — exists between UAs for the provision of the IPMS. It may be
visualised as a standard message header format. P2 is the first of
the so-called 'Pc' protocols or content protocols each of which is
provided along with a new class of UA for particular user
applications;

P3 — exists between a Submission/Delivery Entity supporting a

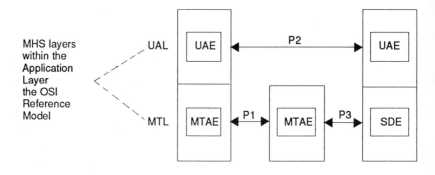

**Figure 4.7   MHS Sub-layering Within the OSI Application
Layer**

single UA remote from its MTA;

P5 — provides a teletex terminal with access to a teletex access unit (TTXAU) and hence to the facilities of the IPMS.

Figure 4.8 indicates how these protocols interact with the elements of a message handling system.

MANAGEMENT DOMAINS WITHIN MHS

After defining the components and services of a message handling system, the next stage is to look at how they will be organised and who will have responsibility for them. In a practical message handling environment the systems will be organised into management domains, and specific organisations or groups will be responsible for their operation and maintenance.

Within the X.400 recommendations two types of management domain are defined:

— Administration Management Domains (ADMDs);

— Private Management Domains (PRMDs).

The ADMDs form the backbone of the messaging infrastructure and are provided by the PTTs. Because of this role they have to provide the majority of optional features defined in the recommendations and in this way they will provide an added value service. These domains will provide connection for individual X.400 users and interconnection between PRMDs. ADMDs have also to provide gateways and access units to other systems and services (see Figure 4.9).

Those systems not under the control of ADMDs will be organised as PRMDs. Any organisation can establish a PRMD in order to provide a service within that organisation. Connection to other PRMDs can be achieved either via an ADMD or via a private connection, eg PSS. The PRMDs will provide a more specialised service which is tailored to the requirements of the individual organisation. PRMDs will, however, be reliant on the ADMDs for interconnection to other messaging services.

Figure 4.10 gives some indication of how these domains might be interconnected both internally within one country and on an international basis between two countries. This is the CCITT view of connections and hence they do not see the potential or need for direct inter-

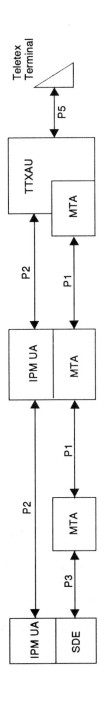

Figure 4.8   Application of P1, P2, P3 and P5 Protocols

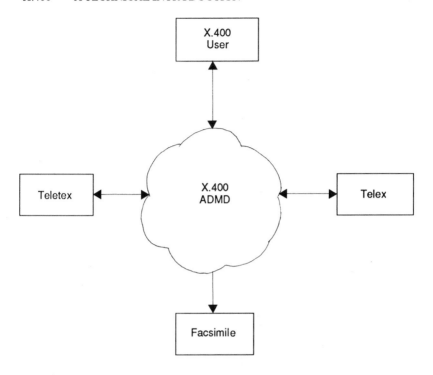

Figure 4.9   Typical Facilities of an ADMD

PRMD links. ISO's MOTIS work does allow such connections and since this work forms the basis of the European CEPT and CEN/CENELEC profiles (see later sections) it is a worthwhile subject to mention.

X.400 NAMING/ADDRESSING

In the past, people have struggled to come to terms with the non-user-friendly addressing schemes employed within electronic messaging systems. In a deliberate move against this potential obstacle to user acceptance it was decided to attempt to rectify this situation in X.400 systems. Hence the concept of the Originator/Recipient (O/R) name and address was introduced. Before discussing these further it is important to specify the difference between an X.400 name and address:

— an O/R name is a descriptive name of the actual human user which is applied to their personal UA and thus specifies an entity

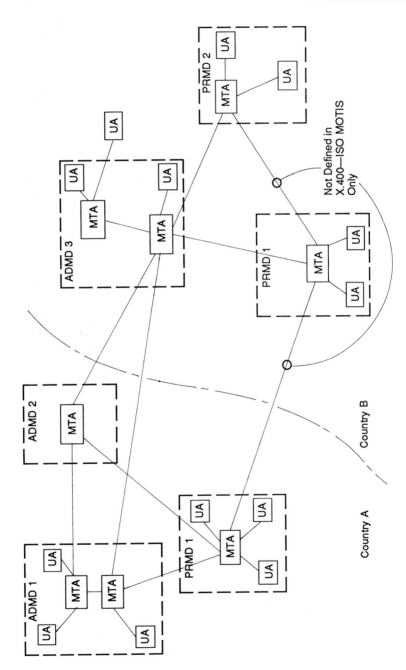

Figure 4.10   Interconnection of Administration and Private Domains

by specifying what it is. Such a name can be used to specify the intended recipient of a message but would require the MTS to perform a directory mapping function in order to determine where the recipient is located, eg which ADMD, PRMD, etc;

— an O/R address is a descriptive name for a UA which has certain characteristics to help the MTS locate the UA within the messaging environment. Every O/R address is an O/R name, but not every O/R name is an O/R address. The use of an O/R address to identify the intended recipient of a message implies that the MTS will not have to use the directory as it has already been given the required information. This feature will be very useful in the early stages of X.400 MHS implementation before the arrival of a full international directory service.

It is the intention that by using the concept of O/R names a message originator will be able to provide a descriptive name for each recipient using information commonly known about that user. Such information might be that which is typically found on a business card, the idea being that just enough information has to be specified in order to uniquely identify the recipient within the messaging environment. Four categories of standard attributes for an O/R name have been identified so far. These categories, along with the attributes supported by MHS protocols are defined below:

Personal Attributes

    personal name

Geographical Attributes

    country name

Organisational Attributes

    organisation name

    organisational unit

Architectural Attributes

    X.121 address

    unique UA identifier (only numeric values)

    administration management domain (ADMD) name

private management domain (PRMD) name

(Note: X.121 is a CCITT recommendation defining an international numbering plan for public data networks.)

This is the list of attributes specified in the 1984 version of X.400. However, it may be expanded to encompass others after future development. The MHS protocols also support Domain Defined Attributes (DDAs) which allows for the transmission of extra addressing information, not standardised within X.400, with the message to allow it to be routed through domains which are not X.400 compatible.

At present there are two forms of O/R name which have been defined within the X.400 recommendations. Form 1, in one of its three variant forms (see below), is meant to identify MHS users only, while Form 2 is intended primarily to identify users of teletex and other telematic services, such as telex and facsimile. These two forms of O/R name are detailed below.

FORM 1

Variant 1

O/R name consists of: Country name

Administration domain name

[Private domain name]

[Personal name]

[Organisation name]

[Organisational unit name]

[Domain defined attributes]

(Note: At least one of the attributes within [ ], with the exception of DDAs, must be selected as well as the first two.)

Variant 2

O/R name consists of: Country name

Administration domain name

UA unique numeric identifier

[Domain defined attributes]

(Note: DDAs optional)

Variant 3

O/R name consists of: Country name

Administration domain name

X.121 address

[Domain defined attributes]

(Note: DDAs optional)

FORM 2

O/R name consists of: X.121 address

[Telematic terminal identifier]

(Note: Telematic terminal identifier is optional)

The eventual aim within X.400 systems is that the majority of directories (implemented in accordance with the CCITT X.500 directory recommendations — see Chapter 3) will be interconnected. Initially, however, it is likely that they will develop in isolation, hence the need for the O/R address concept. Apart from offering a straightforward address-finding facility the directory will also be capable of allowing users to establish distribution lists which can in turn be addressed by a single name.

ROUTEING WITHIN MHS

An X.400 Originator/Recipient (O/R) address is a descriptive name for a UA which has characteristics to help the MTS locate the UA within the messaging environment. A 'route' in X.400 terminology is the path to be taken by a message through the MTS in order to reach the recipient's UA. The O/R address will provide the MTS with the information required to select a route.

The X.400 recommendations only specify the method of routeing between the originator's and recipient's management domains: it was felt that internal routeing within such a domain is outside the scope of the standardisation process. The routeing method specified is an incremental one where each MD along a given route will determine the MTA within the next MD to which the message will be passed. With this method no attempt is made to establish a full route for the message, outside the management domains which are to be used, either by the originating MD or subsequent ones along the route.

DIRECTORY FUNCTIONS WITHIN MHS

Chapter 3 outlined some of the activities currently being undertaken by CCITT and ISO in the field of OSI directories. This work will be used as the basis for establishing a fully interconnected and integrated worldwide MHS directory in the future. The 1988 X.400 recommendations have now fully integrated the concepts of X.500 Directories within the basic MHS model and it is this which will ultimately lead to the use of user-friendly O/R names within an MHS.

Within an MHS there are a number of uses for a directory which need to be considered:

— helping a user with message preparation, ie returning the O/R name of the recipient;

— establishing distribution lists to be used by the user and MTS;

— returning an O/R address when given an O/R name by the MTS, ie used in establishing the required route.

When a user produces a message and indicates the intended recipient by their O/R name the MTS will need to refer to the directory in order to establish the corresponding O/R address. When in possession of the address the required route can be established between the originator and recipient.

In the early days of an X.400 messaging infrastructure it is unlikely that a fully interconnected and integrated directory service will be available. During this period it will be necessary to employ the extensive use of X.400 O/R addresses, with the migration to using the user-friendly O/R names being a gradual process.

INTERFACING WITH LOWER OSI LAYERS

Remote Operations

When a UA and MTA are co-located the communication process between them is interactive, ie UA submitting messages and MTA delivering them. In order that the same situation will apply when a remote UA is communicating with its MTA the interactive P3 protocol has been designed. P3 allows the remote UA to access the MTL facilities provided by its MTA. This access includes the transfer, between the SDE and MTA, of messages and responsibility for them during both the submission and delivery phases.

Remote operations, as defined within recommendation X.410, provide a framework for the specification and implementation of interactive Application Layer protocols such as P3. It specifies a form of remote procedure call mechanism which will allow an operation to be invoked, along with any required arguments. Remote operations also make provision for the results of an operation returned after the completion. In this way the SDE communicates with an MTA, during the submission phase, by remotely invoking operations within it, while during the delivery phase the MTA will invoke operations within the SDE.

When any Application Entity (AE) invokes an operation which is provided by another it does so by entering into an exchange of Operation Protocol Data Units (OPDUs). Remote operations define four such OPDUs:

Invoke — used to invoke some operation and to pass any parameters associated with the operation;

Return Result — returned to the invoker of an operation upon its successful completion along with any results;

Return Error — signals an unsuccessful termination of an operation and passes back diagnostic data;

Reject — 'catch-all' for responding to an unexpected or indecipherable message.

These four basic messages are transferred over the reliable transfer server, as defined within X.410 (considered in the next section).

The Reliable Transfer Server

The reliable transfer server or RTS is that part of an application entity (AE) which is responsible for creating and maintaining associations between an AE and its peers. The AEs in this X.400 context are of course the MTAs and SDEs. The other parts of an AE, in addition to the RTS, are termed the RTS-users (eg MTA and SDE) and are those which drive the application protocol, in this case either P1 or P3.

An association is created between AEs in order to transfer Application Protocol Data Units (APDUs). RTS will create such a connection, maintain it and ensure the reliable transfer of APDUs. There are two forms of association which RTS can create:

Monologue — Unidirectional communications;

Two-Way Alternate — Bidirectional communications but only in one direction at any single instance.

The two-way alternate mode of communication requires the introduction of a 'turn' system whereby an AE can only communicate with its peer in the association when it is in possession of the 'turn'. The RTS and RTS-user interactions are described as a set of service primitives:

Open — Establish an association;

Close — Release an association;

Turn Please — Request for the exchange of the turn;

Turn Give — Exchange of the turn;

Transfer — Reliable transfer of an APDU;

Exception — Indication of transfer failure.

The OPEN and CLOSE primitives are used by an MTA or SDE to establish or terminate, respectively, an association with a peer entity. As indicated earlier the turn commands act effectively as a flag between the AEs to indicate who can communicate when operating in two-way alternate mode. TRANSFER requests the reliable transfer of data while EXCEPTION indicates when a transfer of data has failed.

X.400 — Use of the Lower OSI Layers

The X.400 recommendations define an Application Layer protocol to achieve message handling functionality. In order to transfer data and protocol information between the AEs of an MHS the lower layers of the OSI model must be employed. The requirements of layers 1 to 6 placed upon the OSI model are defined within recommendation X.410 sections 4 and 5, and these are summarised below:

Presentation Layer — Minimal presentation protocol defined within section 4.2.1 of X.410 using X.409 notation;

Session Layer — Extensive use is made of this layer using a subset of the Session Layer BAS (Basic Activity Subset) services;

Transport Layer — X.400 utilises Class 0 which provides the basic network services without additional error detection and recovery;

Network, Data Link and Physical Layers — X.400 makes use of the lower-layer procedures defined by X.25, ie packet switched environments.

Figure 4.11 shows the components of the MHS model including remote operations, the reliable transfer server and the OSI lower layers.

MHS STANDARDS — CCITT 1984 X.400 AND ISO MOTIS

Introduction to Profiles

In order that OSI standards will be applicable to both systems today and those that will be developed in the future, they have been designed so that they have a wide range of options which allow them to cope with many conceivable situations. Because these base documents have little or no pragmatic constraints it is left to the implementors to introduce sensible constraints and select options as they require. Now if each individual implementor were to only add constraints and choose the options which they and they alone required, it would quickly lead to a situation where supposedly open systems were incompatible.

To rectify this potentially troublesome situation various organisations have proposed functional standards or profiles which are derived from the base documents. In this way systems based on the same profile will be compatible.

The purpose of a profile is to effectively take a slice through the base standards so as to make them applicable to real-world situations. A profile may be of one individual base standard, a stack of all seven layers or some number in between these two. The first stage in the development of a profile is to choose which standards are to be used for the individual layers, this forms a 'protocol stack'. This can then be further

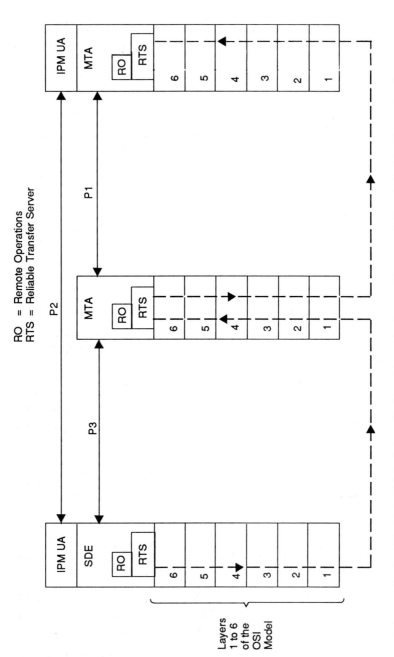

Figure 4.11   The MHS Model Including Remote Operations and the Reliable Transfer Server

refined by the addition of pragmatic constraints (eg setting the maximum permissible length of message in a messaging system) and choosing which options are to be included from the base standards. When this refinement process is complete the resulting document is a profile.

The major uses for the profile are in the procurement and conformance testing of systems. In procurement a purchaser can demand that all their systems should conform to a particular specification in order that compatibilty is maintained. Typically manufacturers claim that their OSI products conform to one or more specific profiles, hence it is possible to see at a glance whether a product is applicable or not. Because a profile, like any other standard, is produced in a human language it is open to ambiguities and misinterpretation. This situation will exist until standards are defined by mathematically-based formal methods which prevent such problems. For the moment, however, the only way to prove conformance to a profile or base standard is to conformance test the implementation. Conformance testing will ensure that a product conforms to a specific profile.

The profiles of the messaging standards are the documents which the user will specify conformance to when procuring products from vendors. Accurate conformance to these profiles is essential if implementations from different vendors are to interwork successfully. The eventual means of checking this conformance will be via the OSI conformance testing services which are now being developed by organisations such as SPAG Services, the Networking Centre and the NCC.

When looking at the implementation of CCITT 1984 X.400 and early ISO MOTIS (extensions for private messaging, ie inter-PRMD) work there are basically three profile documents to be considered:

CEPT A/311—This profile carries European Pre-Norm. status (ENV 41202) and as such will eventually become the standardised messaging profile (PRMD-to-ADMD interworking) for the European Community;

CEN/CENELEC A/3211 — This profile also carries European Pre-Norm. status (ENV 41201) but is complementary to ENV 41202 because it addresses PRMD-to-PRMD interworking;

NBS NBSIR 86-3386-5 — This is a North American profile for

X.400 message handling interworking. It specifies the functions and procedures for PRMD-to-PRMD and PRMD-to-ADMD connections.

The CEPT and CEN/CENELEC work is complementary but these and the corresponding NBS profile, which encompasses the same areas, are not compatible at a detailed level. These problems are recognised, however, and moves are currently under way to try to align these documents as soon as possible.

X.400 SYSTEMS AS FLEXIBLE APPLICATION BASE

The separation of the services provided by an X.400 system is the real key to their underlying flexibility. The generic Message Transfer (MT) service yields the means with which to transfer any kind of information within an X.400 MHS. This service is generally only concerned with transferring information between the sender and recipient, and has no role in the interpretation of that information. The User Services are really the user applications for the MHS, and these provide a service tailored to a specific requirement. An example of this would be message communications between human users; this is of course the Interpersonal Message Service (IPMS) described in an earlier section. As yet this is the only User Service to be standardised.

The IPMS has the ability to carry, in a meaningful way, typical office memos and documents. It allows various attributes to be attached to the information being transferred such as sender/recipient names, subject indication, cross-references, expiry dates, etc. These are the typical types of details used to identify the purpose of documents within an office environment. This functionality is detached from the actual form of the interchange. Hence the IPMS could be transferring an IA5, telex, or facsimile-encoded document or even one employing sections which are encoded in different ways. This flexibility will allow the IPMS to carry any other forms of encoded information which can be registered as body part types when the need for them has been identified.

In the future other User Services to meet specific applications will undoubtedly evolve and all of these will use the generic transfer facilities provided by the MT-service. These new applications will be the key to achieving the maximum utilisation of an X.400 messaging link, and ultimately mean that it will provide a really cost-effective solution to a user's messaging requirements.

Already it is possible to identify some examples of new user applications which are likely to be the subject of future standardisation:

* Electronic Data Interchange (EDI) — The transfer of structured business data which needs to be exchanged during the course of trading operations. Here information contained within invoices, purchase orders, bills of lading, etc is moved electronically between computer processes;

* File Transfer — The transfer of data files within an X.400 MHS;

* Bulletin Board Access — Access to public and private bulletin boards via an MHS.

Further requirements for applications will undoubtedly be indentified as experience with such systems is accumulated. This should not present any problems as the X.400 recommendations were devised with such flexibility in mind. This ultimately leads to the conclusion that an MHS offers an attractive standardised application base which can grow to encompass the ideas of the future.

5 Benefits of Using X.400 Message Handling Systems

INTRODUCTION

Some of the benefits of using the X.400 recommendations as a basis for electronic messaging systems have already been mentioned in earlier chapters of this publication. It is, however, important to identify all of these potential benefits and more, within a single section which will provide a clearer insight into the range of advantages of this approach. This information will be of use to organisations attempting to find the required justification for expenditure on X.400 products. It may also highlight certain implications of X.400 systems which they may not be aware of. Individuals who doubt the worth of such systems should consider the potential benefits listed in this chapter.

Because of the paucity of real user experience with X.400, it is not possible to cite glowing case studies which show the excellence of this technology. Instead reliance must be placed on the potential which this new solution can offer in comparison with existing systems (see Chapter 2). The rationale will be to list the *potential* areas of benefit to be achieved by using X.400 systems; in years to come, as practical experience is gained, these should become apparent in reality.

The benefits to the user are presented in two broad sections:

Benefits to the user

— at an organisational level;

— at the end user level.

BENEFITS TO THE USER

Potential benefits to the users of X.400 systems are wide ranging and encompass far more than can be expected from any other form of electronic messaging media. Part of the reason for this stems from the overriding flexibility which is built into the original recommendations. This flexibility allows X.400 systems to act as a platform for many different applications. The case is helped still further because this is an internationally standardised solution which can be implemented by all countries with security in the knowledge that it is not going to change overnight. Even when change does occur backward compatibility will be ensured and hence equipment investments will be maintained.

Some of the benefits to the user discussed within this section will become apparent right from the outset with an X.400 implementation while others may take a little longer to accrue. This is because some of the benefits will only emerge out of the establishment of a worldwide X.400-based messaging infrastructure which will undoubtedly take some years to develop. Other benefits will only become a reality when an organisation as a whole has migrated to this technology.

BENEFITS TO THE ORGANISATION

Migration towards X.400 MHSs within an organisation can only be achieved with high-level commitment. This kind of commitment can only be reached when the positive effects of such a policy have been clearly identified. It is thus vitally important to look at the benefits of X.400 at an organisation-wide level.

In this context the benefits indicated below are likely to be obtained by the organisation when X.400 is adopted.

Strategic Issues

Maintaining a Market Position

An important aim of most businesses is to establish a good position in the market. This can be done in many ways, one of which is to obtain a reputation for fast, reliable service to customers. In this context an efficient internal and external electronic messaging service is crucial. In the case of a system based on the X.400 recommendations these criteria can be easily achieved. X.400 offers the potential for accessing many existing forms of messaging media. Hence this type of system is not restricted by its own technology boundaries. This means that an X.400 link offers the most suitable means of accessing all the internal branches within an organisation and a suitable basis for direct customer communication, regardless of which systems have to be conversed with. In this way a faster more efficient customer service would be easy to achieve especially when compared to the outdated postal services.

Achieving Competitive Edge

This is another reason for adopting an X.400 solution. If the use of X.400 can produce gains in efficiency and in some cases reductions in cost, these savings can be passed on to the customer in the form of reduced costs, increased speed of service or expanded services for the same costs. Staying ahead is a constant battle: it is not enough to improve efficiency and then sit back and hope to maintain the competitive edge. If competitors choose to adopt the same approach they will soon catch up and possibly take over the leading position. To remain competitive a company must develop and integrate the technology in the best way possible. Because X.400 is a future-proof solution and can act as a basis for many different applications it can grow with an organisation to meet its needs both now and in the future.

Enhanced Image

In a business where it is important to be seen to be keeping up-to-date with technology, it may be possible to enhance the image of a company by using X.400 electronic messaging facilities. With X.400 being based upon the principles of Open Systems Interconnection (OSI), it can be seen that the organisation is looking towards the longer-term benefits to be gained by the use of the technology. It is also possible that a company may wish to publicise its use of message handling via case study material which can be made available to customers. The fact that potential customers could communicate with the organisation via a whole range of electronic media must only serve to impress them.

Improved Communications

Information Access

X.400 can have an important role to play in the integration of many existing communication and information systems within an organisation. Also easier external communications will allow access to many more sources of information. This means that users of the system will have better access to more sources of information. Policy makers should be able to make decisions based upon a better selection of high-quality information. The result of this scenario must surely be improved productivity from the existence of an X.400 system.

Quality of Communications

Both internal and external communications will be vastly improved with the widespread adoption of X.400. Within a short time period it should be possible to see an extensive national and international messaging infrastructure developing. This combined with large-scale penetration of internal X.400 communications within companies will serve to lower long-standing communication barriers. This improved quality of communications will benefit all tiers of the organisation involved in accessing and disseminating information.

System Integration and Interconnection

Because X.400 is an internationally standardised solution it can be used as a general means for the interconnection and integration of systems. For the organisation this means that X.400 can be used as the 'glue' with which to bring together their existing stand-alone means of communications. Integrated office systems, for example, generally offer useful mail facilities to small pockets of users within an organisation. Utilising X.400 gateways between these allows the users of each of the systems to exchange messages, hence, establishing closer working relationships between the various groups.

Purchasing Equipment

Cost Reductions

Because X.400 can offer a single solution to a number of, if not all, a company's messaging requirements there are potential cost savings to be made. This streamlining of communications means that an organisation need only subscribe to one X.400 service rather than to a number of others. Although this service in itself may be more expensive than any one of the previous individual services it is unlikely that it would exceed the total bill for communications; hence there is a cost saving in real terms. These kinds of cost savings can only be achieved when the full potential of X.400 is being used, utilising one link for many applications.

Protection of Investments

X.400 functionality can be added to most existing hardware via a software package which will either allow 'native-mode' or gateway operation. In this way X.400 can be used to protect existing hardware (and sometimes software) investments. Also the addition of X.400 functionality adds real 'value' to existing systems by allowing them to address a far wider user base than before.

Future-proof Solution

X.400 is an internationally agreed solution intended to be the basis of systems for some considerable time. Any future updates of the recommendations will provide backward compatibility paths so as to protect existing investments. In this way systems will become future-proof as they will always be able to communicate with other X.400 systems regardless of age and model.

No More Special Gateways

X.400 should become *the* commonly accepted means for systems interconnection and integration. In this way the need for specialised gateways, between proprietary systems architectures, will be removed. In order to interconnect with any other systems, a manufacturer need only supply a standard X.400 gateway which will meet all the needs of the user. This in itself suggests substantial savings both in the effort and cost required to produce a suitable gateway function.

Freedom of Choice

Freedom to choose equipment on the basis of merit rather than on the basis of whether it will interconnect/integrate with existing purchases is an added benefit of X.400. Because all the equipment from the vendors should conform to the same standard, organisations will no longer have to be constrained in their range of options.

Equal Footing for Suppliers

Suppliers can be dealt with on a more equal footing because the constraints of a particular proprietary system which are currently tolerated can be avoided. The selection of another supplier's equipment will no longer carry with it associated overheads which relate to compatibility. Easy X.400 interworking will reduce the dependence upon one supplier.

Increased Competition Between Suppliers

Because suppliers will have to compete on an equal footing this should mean that equipment prices will start to fall while the quality will increase. This scenario has been shown to be true in many other markets were standards have applied (for example, the de facto personal computer standard of the IBM PC).

X.400 Added Value

Internationally Standardised Solution

X.400 is the first internationally agreed standard for computer-based message handling. This means that its use will expose organisations to benefits which are not apparent with existing forms of messaging media. All around the world organisations, service providers and authorities are starting to adopt X.400 as the basis of their future messaging requirements. These developments will lead to the establishment of a worldwide messaging infrastructure. This will mean that organisations adopting this solution will be able to address, via electronic communications, a far wider marketplace than with any other system.

A Flexible Application Base

It was intended from the outset of their development that the X.400 recommendations would serve as a flexible application base for industry-specific application in addition to simply electronic mail. A brief insight into this flexibility was given in the previous chapter were a number of potential applications were highlighted. For these reasons an organisation's X.400 links must be considered to offer more than merely electronic mail if their true worth is to be achieved. No other form of messaging media can offer this sort of adaptability.

BENEFITS TO THE END USER

So far the benefits of X.400 which can be perceived at an organisational level have been discussed, but there are those which will become apparent to the actual users of systems. These will mainly be in terms of improved ease of use and access. But in time as the usefulness of X.400 comes to fruition it will become a indispensable tool in the everyday work of individuals.

The advantages of X.400 to the end user are:

Easy Access — An X.400 link will give access to many other systems and services, such as telex, other telematic services and computer-based messaging systems/services. For the user this will mean that all messages can be sent from one terminal located on their desk. No longer will it be necessary to learn how to access many different systems, generally achieved with different terminals within separate offices. X.400 will give the user unrivalled easy access to all parts of the globe from a single terminal on their desk.

Information Access — Because X.400 is a means of achieving both easy internal and external interconnection and communications with other systems and individuals it will yield access to much information which was previously either impossible or very difficult to obtain. This improved information access leads to higher personal productivity, less wasted time, and better informed decision making.

Potential for Many Applications — Message handling systems have been defined in such a way that they may act as a basis for many applications, other than just electronic mail. Ultimately this will mean that users will be able to operate many applications from a single terminal.

Advantages Over the Telephone — These are benefits which can be incurred with the use of any electronic text communication service, but in the case of X.400 its impact will be substantially wider and hence its benefits more apparent:

— easy access across time zones;

— no need for 'telephone-tag' scenario;

— hard copy of a message removes the need for rough, sometimes inaccurate, telephone notes;

— text based messages are more accurate and well defined;

— fear of telephone communications is eliminated.

Less Reliance on Slow Postal Services — In comparison with other forms of electronic messaging X.400 will provide access to a far wider audience. This will mean that users will place much less reliance on outdated and slow postal services as their means of every-day communication.

SUMMARY

It should now be clear that the adoption of the X.400 solution can hold the key to a number of benefits for both the organisation and the user. Certainly the suppliers have been quick to realise the potential of this new standard and already a substantial number of products and services have appeared in the marketplace. This supplier migration is doubly interesting because of its scope. It now seems that all the major First World countries around the globe are carrying out some form of X.400 implementation activity. This points the way, ultimately, towards the establishment of the worldwide messaging infrastructure.

The final element which is required is that of user adoption of X.400. This has been slow up to now but is likely to increase rapidly as the X.400 services become readily available. When this has been achieved the true benefits of this approach should become really apparent.

6 1988 X.400 — The Future

INTRODUCTION

It has already been stated within this publication that the original CCITT X.400 recommendations, as ratified in 1984, contained a number of errors and ambiguities. Although these were effectively dealt with in the subsequent *Implementors' Guides* there was a clear need for a 'clean' set of base documents for this important topic area.

In addition to this, the concepts of OSI Application Layer architecture, which were not fully established when the original X.400 work was being completed, had since been finalised. From this it was clear that the existing MHS architecture model and sub-layering modelling techniques did not fit within this new framework.

Certain areas of the 1984 recommendations were left for further study at the time of ratification, this was generally due to a lack of time within the study period. Further, since the ratification of the original messaging work a number of associated topics were being defined, such as Security and Directory functions, which have a bearing on this area of development. In addition to these, the requirement for a number of extra MHS facilities had also been identified, such as a message store and connection to physical delivery services.

All of the above reasons and others meant that the CCITT would continue their work on the message handling topic throughout the subsequent study period after the ratification of the original documents. The goals of this development were to:

— produce a set of documents which were as near free from errors and ambiguities as was possible;

— encompass the new concepts of the OSI Application Layer;

— retain backward compatibility with existing 1984 X.400 implementations;

— include some new features and facilities which were deemed to be of use.

This development process culminated in the production of the 1988 X.400 drafts, completed in mid-1988, and proceeded through an editorial phase and voting period before ratification in late-1988.

The original ISO Message Oriented Text Interchange System (MOTIS) work was an adoption and expansion of the CCITT 1984 recommendations. The development of X.400 by ISO was largely to allow message handling facilities to encompass private messaging requirements. MOTIS provided the user with the option of linking private messaging domains directly, hence yielding the possibility of forming private messaging systems, ie PRMD-to-PRMD interworking.

In Europe it is important to take account of this work as it has formed the basis of the major European messaging profiles from CEPT and CEN CENELEC. Since these profiles are now classified as European Pre-Norms, and will eventually reach full European Norm status, they have special significance for European procurement situations. This is now especially apparent for Government Departments and Local Authorities after the recent EEC Decision (87/95) which will apply in most IT procurement situations.

In the light of this it is perhaps surprising that MOTIS was not progressed through to full International Standard (IS) status. Rather it was abandoned at the Draft-IS (DIS) stage in early 1987. This decision was taken in the light of the forthcoming 1988 work from CCITT and the amount of time and effort required to bring the original ISO documents through to full IS status. ISO preferred to discard the original work and to adopt the current drafts from the CCITT and to work with them to achieve a standard and recommendation with common text.

Although the new CCITT recommendations for message handling were only ratified during late-1988 for subsequent publication in early-1989 it is likely that their adoption will be more rapid than the original work. In 1984 there was little experience of X.400 concepts and hence some inertia to overcome before their implementation. The later half of

1987 and early part of 1988 have really been the take-off period for 1984 X.400 products and services, a time lapse between ratification and implementation of approximately four years. 1988 X.400, however is emerging when initial experience with messaging concepts has been gained; hence there is no inertia to overcome. Further, the added value, in terms of features and facilities, which these new recommendations offer will be the real incentive in the move towards their implementation. It is possible to conceive of the first products appearing on the market by late-1990/early-1991, ie a substantially shorter time-scale.

It must be emphasised that the situation with this updated recommendation in no way parallels developments with others, such as X.25. The X.25 recommendations have been updated on three occasions, the latest of which was in 1984. This later version of X.25 offers no real added value in terms of features or flexibility. Hence, there is a lack of incentive to implement them when earlier versions appear to be adequate. 1988 X.400 is a definite improvement on what 1984 can offer. Hence, its adoption would seem to be assured.

STRUCTURE OF 1988 X.400 RECOMMENDATIONS

The revised structure of the 1988 X.400 recommendations is as follows:

X.400 — System and Service Overview A good introductory document which looks at the general concepts of MHSs including short sections on the many new features which have been included.

X.402 — Overall Architecture A fairly detailed introduction to the concepts of 1988 messaging characteristics, the MHS model and how this relates to the new Application Layer architecture.

X.403 — Conformance Testing (for X.400-1984) This recommendation describes the test methods, criteria and notation to be used for conformance testing 1984 MHSs (as supplemented by Version-5 of the *Implementors' Guides*). This basically acts as a definition of what a 1988 system will interwork with.

X.407 — Abstract Service Definition Conventions Specification of the services to be provided in an MHS environment in an abstract form. This effectively separates the description of the service from its concrete realisation.

X.408 — Encoded Information Type Conversion Rules Similar to the 1984 recommendation but with the items classified for 'further study' now complete. It specifies the algorithms which the MHS will employ when converting between different encoded information types, eg IA5 to telex.

X.411 — Message Transfer System: Abstract Service Definitions and Procedures Defines the abstract service provided by the MTS and specifies the procedures to be performed by MTAs to ensure correct distributed operation of the MTS.

X.413 — Message Store: Abstract Service Definition This recommendation defines the procedures for using the Message Store (MS) and indirect message submission through the MS to the MTS.

X.419 — Protocol Specifications This details the remote UA access protocol (P3), the MS access protocol (P7) and the message transfer protocol (P1).

X.420 — Interpersonal Messaging System This deals with the Interpersonal Messaging Service (IPMS), a user service for the exchange of messages between human users. The major difference between this recommendation and the 1984 version is the lack of the Simple Formattable Document (SFD) specification which has now been dropped.

The correlation between the CCITT MHS and ISO MOTIS numbering schemes is shown in Table 6.1.

Other related ISO and CCITT recommendations are:

ISO 7498 : X.200 — OSI: Basic Reference Model;

ISO 8824 : X.208 — OSI: Specification of ASN1;

ISO 8649/2 : X.217 — OSI: Association Control: Service Definition;

ISO 9066/1 : X.217 — OSI: Reliable Transfer: Model and Service Definition;

ISO 9072/1 : X.219 — OSI: Remote Operations: Model, Notation and Service Definition.

CCITT	ISO
X.400	10021-1
X.402	10021-2
X.407	10021-3
X.411	10021-4
X.413	10021-5
X.419	10021-6
X.420	10021-7

Table 6.1

1988 MHS FUNCTIONAL MODEL

When compared to the 1984 MHS, the new functional model (see Figure 6.1) has basically three additional features:

— the concept of access units has been formally introduced, although these will be or are already offered with 1984-based X.400 services, to allow MHS users to communicate with telex and telematic services. The rules for coded information conversion, which were left for further study in the first issue of the recommendations, have now been fully defined and hence it is now possible to standardise the conversion of message contents in order to achieve a coherent strategy on a worldwide basis.

— access to physical delivery (PD) services was identified as another means of achieving a critical mass situation with X.400 systems, hence, it was decided to introduce this into the new recommendations. An MHS user will be able to have a message delivered to its recipient by sending it to the PD access unit

(PDAU) which will produce a hard copy with an envelope addressed with the electronic addressing information provided by the originator. A special type of Postal O/R address has been introduced which contains the typical sort of information to be found in a normal postal address along with additional information about the type of service to be used, eg postal or courier.

— the final feature to be highlighted from the functional model is the potential for the inclusion of a message store (MS) between an MTA and its remote UA. This overcomes the problems outlined in Chapter 4 with respect to 1984 remote UA arrangements.

CHANGES TO 1988 MHS MODEL

Late in the 1985/88 study period it was decided that the then-current MHS drafts should receive a major re-draft to attempt to embody the

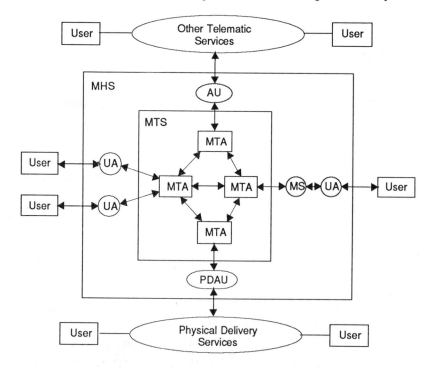

Figure 6.1   The 1988 MHS Functional Model

newly established concepts of OSI Application Layer architecture. This re-draft involved the removal of 1984 sub-layering concepts and the introduction of blocks of functionality which can be combined in a number of ways to provide different forms of connection dependent upon the operation to be accomplished.

In Chapter 3 the OSI Application Layer structure was reviewed in some detail. This indicated that communication between application processes is represented in terms of an association between application entities (AEs) residing within the Application Layer. Each AE is a grouping of functionality which has been combined in order to achieve a specific task such as message handling or file transfer. The functionality of an AE is subdivided into a set of one or more application service elements (ASEs). Any interaction between AEs is described in terms of their use of services provided by ASEs. Within the 1988 X.400 recommendations the following ASEs are described:

Access to the Message Transfer (MT) service

— Message Submission Service Element (MSSE);

— Message Delivery Service Element (MDSE);

— Message Administration Service Element (MASE).

Access to the Message Store (MS) service

— Message Submission Service Element (MSSE);

— Message Retrieval Service Element (MRSE);

— Message Administration Service Element (MASE).

These ASEs are concerned with the operations of message transfer/ receipt and as such require the support of other ASEs which are concerned with connection establishment/release and information transfer in an OSI environment. Such facilities are provided by the following ASEs:

— Remote Operations Service Element (ROSE) — Supporting interactive request/reply operation within the MHS model;

— Reliable Transfer Service Element (RTSE) — Supporting the reliable transfer of application data;

— Association Control Service Element (ACSE) — Supporting the establishment and release of connections between a pair of AEs.

In order to observe how these ASEs interact, refer to Figure 6.2 which shows the permissible application-contexts. There are three basic ones to consider: a remote user accessing the MTS; a user accessing the MS; and the basic message transfer operation.

MESSAGE STORE

One of the obvious omissions from the CCITT's 1984 recommendations was the lack of a message store (MS) between an MTA and its remote UA/s. This meant that if the remote UA was 'off-line' for a period of time it could be 'flooded' with messages at the log-on stage. If at this time there is insufficient information storage available then message loss can result. The 1988 X.400 recommendations include MS concepts so allowing the remote UAs to be 'off-line' without the danger of message loss when the UA becomes 'live' again.

The functionality defined for the message store can be summarised as follows:

— one MS acts on behalf of one user (ie one O/R address);

— when subscribing to an MS all messages destined for the UA are delivered to the MS. When a message is delivered to an MS the role of the MTS in the transfer process is complete;

— it is possible to request an alert when a certain message arrives;

— message submission from the UA to its MTA, via the MS, is transparent;

— users are provided with general message management facilities such as selective message retrieval, delete and list.

SECURE MESSAGING

The requirement for security features within a computer-based MHS was identified as another area for the 1988 X.400 recommendations to address. The features which have been defined are certainly extensive and can be classified into three broad classes:

— Originator to Recipient — These are features which are operated on an end-to-end basis and do not require the use of MTS security features;

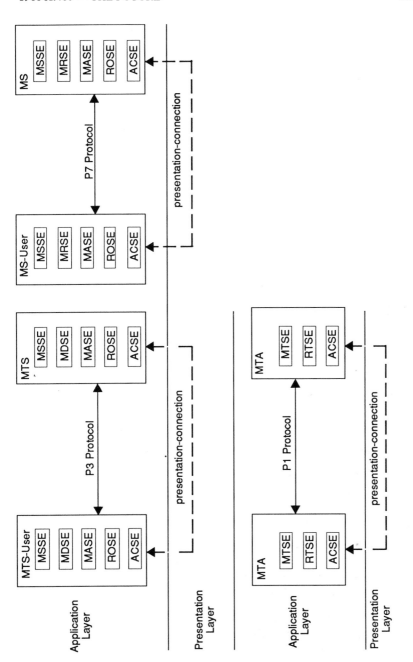

Figure 6.2   MHS Application Contexts

— Message Transfer System — Some features are provided by the MTS and hence require the UA to interact with it in order to invoke them. Note that this also implies that the MTA being used is equipped with the appropriate security functionality;

— UA, MS and MTA — Some security features apply not only to the UA and MTA but also to the MS and in particular to the status of certain messages held therein.

A full list of the 1988 security capabilities is contained in Figure 6.3 which is an extract from draft text of the 1988 version of the X.400 recommendation.

MHS DIRECTORY ACCESS

The need for a fully integrated directory service within an MHS environment is unquestionable. It is the means by which users can find addressing information and the source of the information which the MTS will use to establish a route for a message across the MHS between the originator and recipient. It was this goal that led the CCITT to commence work on their X.500 series of Directory Service recommendations. ISO have also adopted this work (ISO 9594), because of their global OSI addressing needs, and undertaken a joint development process with the CCITT which parallels the approach taken with the 1988 MHS/MOTIS work.

The principles of the directory service were adequately covered within Chapter 3. Hence, these do not need further explanation. These principles have been adopted, and the directory service fully integrated with the MHS in the 1988 X.400 recommendations (see Figure 6.4). This diagram indicates that both the UA and the MTA can have directory access by having Directory User Agent (DUA) functionality built into them. This allows access to the Directory Service Agents (DSAs) which hold the MHS addressing information. A user creates a message in cooperation with a UA and when addressing information is required it can be retrieved from this point by allowing directory access. When a message is submitted to the MTS for delivery, the system will access the directory for the information required to establish its route across the MHS.

SUMMARY

It is clear that the 1988 versions of the X.400 recommendations do

Message Origin Authentication. Enables the recipient, or any MTA through which the message passes, to authenticate the identity of the originator of a message.

Report Origin Authentication. Allows the originator to authenticate the origin of a delivery/non delivery report.

Probe Origin Authentication. Enables any MTA through which the probe passes, to authenticate the origin of the probe.

Proof of Delivery. Enables the originator of a message to authenticate the delivered message and its content, and the identity of the recipient(s).

Proof of Submission. Enables the originator of a message to authenticate that the message was submitted to the MTS for delivery to the originally specified recipient(s).

Secure Access Management. Provides for authentication between adjacent components, and the setting up of a security context.

Content Integrity. Enables the recipient to verify that the original content of a message has not been modified.

Content Confidentiality. Prevents the unauthorised disclosure of the content of a message to a party other than the intended recipient.

Message Flow Confidentiality. Allows the originator of a message to conceal the message flow through MHS.

Message Sequence Integrity. Allows the originator to provide to a recipient proof that the sequence of messages has been preserved.

Non Repudiation of Origin. Provides the recipient(s) of a message with proof of origin of the message and its content which will protect against any attempt by the originator to falsely deny sending the message or its content.

Non Repudiation of Delivery. Provides the originator of a message with proof of delivery of the message which will protect against any attempt by the recipient(s) to falsely deny receiving the message or its content.

Non Repudiation of Submission. Provides the originator of a message with proof of submission of the message, which will protect against any attempt by the MTS to falsely deny that the message was submitted for delivery to the originally specified recipient(s).

Message Security Labelling. Provides a capability to categorise a message, indicating its sensitivity, which determines the handling of a message in line with the security policy in force.

Figure 6.3   1988 MHS Security Capabilities

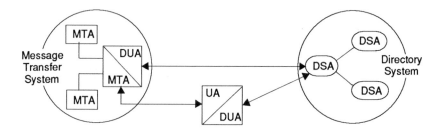

**Figure 6.4   The Functional Model of MHS Directory
Interworking**

indeed offer additional functionality. The standardisation of the Message Store, extensive security and directory access facilities mean that ultimately this route will be the most useful. In addition the backward compatibility built into 1988 will ensure that existing investments are maintained wherever possible. This is an important point as it will be some time before implementations based on this new work will be commercially available. Hence users should not be constrained by only looking at the implementations of the 1988 recommendations. Rather, it is better to gain experience with current 1984 implementations as the potential benefits (see Chapter 5) of these are sufficient to justify their purchase. Considering all factors it does indeed appear that the 1988 X.400 recommendations will be 'the future' for message handling, but this does not devalue any of the current work being carried out on X.400 implementation.

7 Summary

Today's means of electronic messaging offer insufficient scope to meet the needs of present, let alone future, business environments. Such systems are limited either by their horizon, not being implemented in accordance with an international standard, or by the flexibility and facilities which they can provide. The user bases of these existing services have developed in isolation with the lack of a coherent strategy for their interconnection and integration. In many cases organisations have to resort to a number of types of messaging media to enable them to address all their business and trade partners. This mode of operation is quite obviously inefficient and requires the user to be familiar with the operation of a number of systems.

So far then, the use of electronic messaging can only be considered to be at an embryonic stage. This form of communication can potentially offer substantial advantages over the traditional telephone and postal services. But as yet it has not rivalled them seriously in terms of the volume of information carried. In order to make electronic messaging a more attractive option a substantial development of the existing infrastructure is required. Standardised access between all systems is one issue; in this way, the existing user bases will be joined together and hence begin to perform as a coherent whole. Another requirement is that any new systems are developed to a recognised international standard. This will allow access between existing systems and a secure basis for all future messaging activity in its potentially varied forms. It is only by this process of development that electronic messaging will emerge from its embryonic stage to achieve the maturity it actually deserves.

The CCITT's X.400 Message Handling System (MHS) recommendations provide an internationally standardised solution with which it is possible to achieve these goals. Preservation of existing investments was always one of the main aims of this approach. In this way X.400 can act as a gateway between existing systems while providing the kind of flexible base required for future systems. Hence, the horizons of X.400 systems are not limited by the lack of standardisation or flexibility. The interconnection and integration of existing systems along with new ones will mean that investments are maintained, in fact their value should be increased substantially because of the potential to access whole new communities. Electronic messaging should eventually emerge as *the* major form of intra-/inter-business communications allowing real improvements in the efficiency and speed of information transfer.

To consider systems based upon the X.400 recommendations merely as being useful for electronic mail is to seriously underestimate their capabilities. Message Handling Systems can transfer any type of structured or unstructured data, and hence provide a flexible base for user-specific applications. Various examples of these have been discussed in this book but it must be emphasised that the potential for others is only limited by the imagination of the user. As their experience with X.400 develops, they will start to see how it can impact on other parts of their business operation rather than just electronic mail. Ultimately it is possible to foresee a single X.400 connection providing the user with access to many applications.

Open Systems Interconnection (OSI), the underlying architecture upon which an X.400 message handling application is based, is likely to have a significant part to play in all aspects of computer communications. The OSI framework provides the means by which data is reliably routed and transferred between user applications. Conformance to OSI standards is going to become more and more important in the future, indeed it has already assumed some importance to certain groups. A recent EEC decision requires that all public authorities specify adherence to OSI standards in systems costing over 100,000 ECUs (European Currency Units) or approximately £70,000 at today's values. Exemption is possible on a number of counts, including lack of suitable OSI products and on a cost justification basis. However, these arguments will become less and less feasible in the near future. The emergence of X.400 as the first readily available OSI-based user application is beginning to raise the awareness of the OSI communications solution and should ultimately have a significant part to play in its adoption.

A recent NCC survey of UK local authorities attempted to assess the level of awareness within this group which is affected by the EEC decision. The results (see Figure 7.1) show a generally low level of awareness of OSI communication although the PSS services and private X.25 communication systems already play a significant role. It was, however, interesting to see X.400 message handling systems already playing a notable part in the plans of local authorities. At present approximately 3% of the sample (over 100 responses) were actually using X.400 with a further 19% planning its use in the near future. This would seem to indicate that the potential of message handling systems is already influencing the user's view of OSI as a basis for future communications requirements.

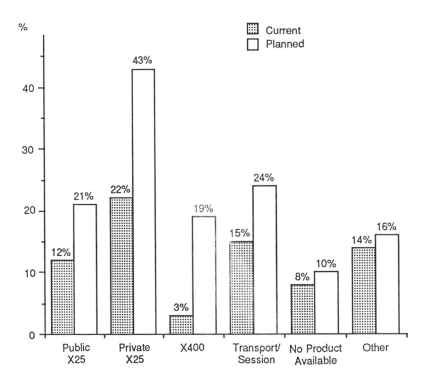

Figure 7.1   NCC Local Authority OSI Survey Results

One of the major attractions of OSI-based applications, such as message handling, to the user is the removal of the need for single supplier allegiance. In the past organisations have been constrained in their choices of new products by existing equipment investments. Lack of interconnection and integration between products from different manufacturers was one of the major spurs in the development of OSI. Purchasing products conforming to an internationally standardised communication architecture removes the worries of interconnection and integration. In this way the user can make procurement decisions based purely on cost and performance criteria — a much more desirable situation. The introduction of international standards also means that vendors have to compete in an open market. This situation leads to increased competition between them which eventually means better products at lower prices for the user.

Already a large number of X.400 products are beginning to appear, and some of these have been exhibited at the recent spate of interworking demonstrations. Such events are beginning to convey the message that X.400 has really arrived; it is no longer just another international standard. Both ceBIT 87 and Telecom 87, for example, marked a significant step forward for X.400 because they demonstrated the interworking of 'real' products and services rather than just prototypes. X.400 services, the messaging platforms for individual countries, are starting to emerge. Telecom 87, in particular, illustrated the potential for the interconnection of such services on a global basis. This event has given a substantial pointer towards the first stages in the development of a worldwide X.400 messaging infrastructure.

Overall then it appears that X.400 offers great potential as the basis for a future worldwide messaging infrastructure. It is a flexible solution which is not limited to a narrow range of applications or information types. In addition it is an internationally standardised solution which can be considered as future-proof, and hence a stable basis for implementation. All of these factors support the view that X.400 will have a crucial role to play in moving and accessing all the types of information used within business environments.

Appendix A
Glossary

This glossary defines many of the message handling terms used throughout this book. It also defines many of the commonly used terms from the general area of communication based on the principles of OSI.

Abstract and Transfer Syntaxes — These reside within the OSI Presentation Layer but are concerned with the specification of the structure and contents of application data. An abstract syntax will provide for the definition of application data types and indicate their required structure. When the transfer syntax is applied to the resultant abstract syntax it will convert it into a form suitable for transmission.

ACSE — Associated Control Service Elements. Functions defined as part of the Application Layer, common to a number of application services (DIS 8649/2 and DIS 8650/2).

Access Unit (AU) — In the case of a Message Handling System (MHS) these are the units which will act as gateways to other communication systems that are not based upon X.400; examples of these would be telex and other telematic services such as facsimile and teletex.

AD — Addendum to ISO Full International Standard (eg ISO 8348/ ADDI).

Administration — In the context of MHSs CCITT define an administration as either a PTT (Post, Telephone and Telecommunications operator) or a Recognised Private Operating Agency (RPOA).

Administration Management Domain (ADMD) — These are management domains within an MHS which are under the control of an administration. These will provide the X.400 message handling services for users.

Application-entity — The part of an application process which concerns OSI.

Application Process — An OSI term to describe a user of the OSI infrastructure — whether it be an application program, a human operator or a process control device.

Architecture — A framework for a computer or communications system which defines its functions, interfaces and procedures.

ASCII — The American Standard Code for Information Interchange; based on the ISO 7-bit data code, usually transmitted in 8-bit characters incorporating a parity bit.

ASN.1 — Abstract Syntax Notation One; part of the OSI Presentation Layer Standards (ISO 8824 and ISO 8825).

Asynchronous — Simple communication of data character by character which generally relies on a host for screen formatting.

BAS — Basic Activity Subset; one of the defined subsets of the Session Layer (ISO 8326 and ISO 8327).

BCS — Basic Combined Subset; one of the defined subsets of the Session Layer.

Bit — Binary digit; when referred to in bits per second (or bps) it indicates the transmission rate of a communications link.

Body — In the context of an MT-service this is the component of a message which contains the originator information which is to be communicated.

Body Part — The body of an X.400 message may consist of several constituent sections which are termed body parts.

Byte — A group of eight binary digits which form the basic binary word used to convey information; for example, the bit sequence within a byte can represent the code for a letter or a number (ie ASCII).

CASE — Common Application Service Elements; a set of OSI layer 7 standards which provide common functions for the application programs and other Application Layer standards (DIS 8649 and DIS 8650). No longer widely used — more likely to see ACSE and CCR.

CCR — Commitment, concurrency and recovery; functions defined as part of CASE which are common to many specific applications (DIS 8649/3 and DIS 8650/3).

Connectionless Service — Where no permanent connection can be assumed, and no connection establishment takes place prior to communication.

Connection-oriented Service — Where a permanent connection (either logical or physical) exists for the duration of the communication.

Content — In the context of an X.400 MHS this is the information within a message that the MTS will neither examine nor modify, unless conversion is specifically requested, during its transfer of the message between originator and recipient.

Content Type — This is an identifier placed upon the message transfer envelope which specifically defines the type of contents.

Conversion — Message handling conversion is the situation where the MTS will perform a transformation from one encode information type to another in order that the message can be delivered into the recipient's environment.

CSMA/CD — Carrier Sense Multiple Access/Collision Detection, one of the major classes of low-level network technology, and a method of preventing data corruption used mainly for local area networks (LANs). ETHERNET is an example of this type. It is specified in OSI Standard ISO 8802/3, based upon the work of IEEE committee 802.3.

Delivery — When the recipient's MTA receives a message it will deliver it to the recipient's UA in a delivery envelope containing the message body and any header material.

Draft Addendum (DAD) — A draft addendum to an ISO Full International Standard.

Directory — Directory is a generic name which implies a collection of open systems which are cooperating to provide directory services to both users (eg finding a recipient's X.400 O/R names) and to systems (eg network addresses to establish the required route).

Directory Name — The name of an entry within a directory.

Directory System Agent (DSA) — These are the main building blocks of an OSI directory which provide access to directory information for both DUAs and other DSAs.

Directory User Agent (DUA) — The user interface to the directory service which will employ its own DSA, and others where necessary, to locate the required directory information for the user.

Distribution List — In an MHS a distribution list is a specified list of users, and their associated address information, to whom an originator can send a message by invoking the list via its directory name.

Domain Defined Attributes (DDA) — These are X.400 O/R name attributes which are specific to the management domain within which the recipient exists. These may allow a message to be routed to systems outside the MHS.

DIS — Draft International Standard; final stage prior to becoming a Full International Standard from ISO (eg DIS 7942).

DP — Draft Proposal; a proposed standard from ISO at the first stage of development but with some technical stability (eg DP 8613).

EN — European Norme; a standard within the European Community.

End-system — Strictly used in the OSI context to define an 'open system' that can communicate with other end-systems via OSI protocols, as distinct from a RELAY or GATEWAY that performs an intermediate routeing function.

Entity — Active element within one OSI layer which employs the services of the next lower layer to communicate with a peer entity.

ENV — European pre-standards from CEN/CENELEC/CEPT.

Envelope — A message handling envelope will contain the information which the originator has specified for transfer and details to be used during the submission, relay and delivery phases of message transport across the MHS.

Explicit Conversion — This is a specific form of MTS content conversion in which the originator has specified both the initial and final encoded information types.

Electronic Data Interchange (EDI) — EDI is the electronic transfer of structured business data, directly between computer systems, in accordance with agreed message standards.

Electronic Data Interchange for Administration, Commerce and Transport (EDIFACT) ISO 9735 — EDIFACT is the recently agreed ISO Full International Standard for EDI which defines the syntax to be used during such transactions.

ESPRIT — European Strategic Programme of Research in Information Technology.

Ethernet — A type of Local Area Network based upon CSMA/CD technology — originally developed by DEC, Intel and Xerox.

EurOSInet — European demonstration of an OSI network by a number of leading vendors.

Frame — A unit of HDLC. A sequence of bits which make up a valid message, containing flags, control field, address field, a frame check sequence, and optionally an information field.

FTAM — File Transfer, Access and Management — one of the protocols being developed for the OSI Application Layer; specified in DIS 8571.

Functional Standards — Indentified 'stacks' of base standards to allow the construction of interworking products.

Gateway — An intermediate system in the communication between two or more END-SYSTEMS which are not directly linked and/or observe different PROTOCOLS (eg between an OSI system and a non-OSI system).

GOSIP — Government OSI Profile; work from the CCTA to develop functional standards for the UK.

Header — A component of an Interpersonal (IP) message which defines the control information which characterises the IP-Messaging Service (IPMS).

HDLC — High-level Data Link Control; a standard for frame structures in connection with data communications protocols, at the Data Link Layer (ISO 3309 is concerned with HDLC frame structure).

Host — A computer system on which applications can be executed and which also provides a service to connected users and devices.

IGOSINET — An OSI network organised by IGOSIS. Purpose is to encourage interworking.

IGOSIS — Implementors' Groups for OSI; organised by ITSU of the DTI.

Interpersonal Messaging Service (IPMS) — This service provides human users with facilities to exchange messages via the MHS. It is similar in many ways to existing electronic mail facilities and allows many types of message content to be exchanged, eg from ASCII to Group 3 facsimile images.

IP-Message — The content of a message in the IPMS, ie the user information to be transferred.

Implicit Conversion — This is another form of MHS content conversion in which the MTS will perform the selection of initial and final encoded information types.

IPM Protocol (P2) — The protocol which operates between IPM-User Agents for the provision of the IPMS.

IPM-User Agent — The IPM-UAs are the specific class of UAs which provide the IPMS. Further classes of UAs will eventually be defined to meet other specific user requirements.

Interconnection — A term often used to define a lesser level than full INTERWORKING, such that two computer systems can communicate and exchange data but without consideration of how the dialogue between Application Processes is controlled or how the data is presented and recognised.

Interworking — Ultimately, the achievement of proper and effective communication or 'linking' between different Application Processes or programs and data; may be on different systems from different manufacturers, remote from each other and connected by some transmission medium or network.

IS — International Standard; fully agreed and published ISO standard (eg ISO 7498).

Integrated Services Digital Network (ISDN) — ISDN services will emerge out of the current modernisation, from analogue to digital, of the world's telephone networks. They can provide integrated voice and data facilities by extending the digital connection right to the customer's premises. The CCITT are currently developing their I-series of recommendations for ISDN which will be OSI-compatible at the lower three layers of the model.

IT — Information Technology; a term used to encompass the methods and techniques used in information handling and retrieval by automatic means, including computing, telecommunications and office systems.

JTM — Job Transfer and Manipulation; one of the protocols being developed for the OSI Application Layer for activating and controlling remote processing (DP 8831 and 8832).

Kernal — Service elements within the Session Layer which are necessary to set up and close down a connection; part of ISO 8326 and ISO 8327. Also used to describe basic elements of CASE (DIS 8649 and DIS 8650).

LAN — Local Area Network; spans a limited geographical area (usually a building or a site) and interconnects a variety of computers and other devices, usually at very high data rates.

LAPB — Link Access Procedure Balanced; a variant of HDLC used between peer systems, which is the basis for layer two of X.25 (as an example of standards in this area, ISO 7776 is concerned with X.25 LAPB compatible DTE Data Link Procedure).

Layer — In the OSI Reference Model, used to define a discrete level of function within a communications context with a defined SERVICE interface — alternative Protocols for a particular layer should then be interchangeable without impact on adjoining layers.

MAP — Manufacturing Automation Protocol; initiated by General Motors in order to force suppliers to adhere to a prescribed set of OSI-based standards.

Medium — The physical component of a network that interlinks devices and provides the pathway over which data can be conveyed. Examples include coaxial cable and optical cable.

Message Handling System (MHS) — The generic term used to describe an X.400 message handling environment which is a collection of interconnected MTAs and UAs.

Message Oriented Text Interchange Systems (MOTIS) — The ISO name for both the message handling environment and their draft messaging standards; a superset of X.400 expanding the message handling functionality into the private domain. The original MOTIS work based upon the 1984 X.400 recommendations was recently abandoned but the name has been resurrected for their joint work with the CCITT on the 1988 version of message handling.

Message Transfer Agents (MTAs) — These are the main store-and-forward building blocks of an MHS, they provide the message transfer and relay capabilities which are used by the UAs.

Message Transfer — This term is used to describe the store-and-forward message delivery process which is employed within an MHS, this movement of the message is between the MTAs of the system.

Management Domain — In order that the whole message handling environment can be subdivided into sections for administrative and maintenance purposes the concept of management domains (MDs) has been included within the X.400 recommendations. There are two types of MD which have been defined in X.400: ADMD; PRMD.

Management Domain Name — In order to enable systems to route messages within the MHS each MD has a name which can be quoted in the address field of a message.

Message Transfer Layer (MTL) — This concept is strictly 1984 X.400 as it is not in line with current concepts of the OSI Application Layer. The MTL is the lower of two sub-layers within the Application Layer which provides the MT-service, the upper of these two layers is the User Agent Layer or UAL.

Message Transfer (MT) Service — The MT-service is that service within the MHS which provides the basic store-and-forward message transfer capability, in general it is not concerned with the contents of a message unless a content conversion is to be performed.

Message Transfer Protocol (P1) — P1 is the protocol which provides the MT-service by enveloping the content and header before it is transferred across the message handling environment.

Message Transfer System (MTS) — The MTS is the interconnection of all MTAs which provide the message transfer service elements.

Message Store (MS) — This is a new concept from the 1988 X.400 recommendations which provides an intermediary message store 'buffer' between a remote UA and its MTA.

Message — In the context of an MHS this is the unit of information transferred by the MTS and consists of both the envelope and its contents.

Multiplexing — The carrying of more than one data stream over the same connection (apparently) simultaneously.

Network — A collection of equipment and/or transmission facilities for communication between computer systems (whether a single dedicated link, line, dial-up PSTN (telephone) line, public or private data network (PDN), satellite link, etc) more correctly in the OSI context used to define the achievement of end-to-end communication between End-systems, however accomplished.

Network Architectures — A generic term for the layered approach which individual vendors take towards development of their communications and applications products. Examples include IBM and SNA, DEC and DNA, Bull and DSA, and ICL and IPA.

Network Layer — Level three of the OSI model. It is the means of establishing connections across a network such that it then becomes possible for transport entities to communicate.

Node — A focal point within a Network at which information about a network entity is considered to be located.

NSAP — Network Service Access Point; the Service Access Point which allows entities within Network and Transport Layers to interact. It is situated upon the Network Layer boundary. It is located by its address which is the subject of ISO 8348/ADD2.

ODA — Office Document Architecture (DP8613/2); a proposed architectural structure for an office document which allows its logical and layout structure to be defined in an unambiguous manner.

ODIF — Office Document Interchange Format (DP8613/5); a structure for interchange of complex office documents.

ONA — Open Network Architecture; a set of OSI profiles from British Telecom which are to be supported when connected to BT services.

OSI — Open Systems Interconnection; a term which is used to describe the area of work concerned with vendor independent standardisation, largely carried out under the guidance of ISO.

OSI Gateway — A method of providing access to an OSI network from a non-OSI system by mapping the sets of protocols together.

OSI Reference Model — Seven-layer model defined by an ISO sub-committee as a framework around which an Open Systems Architecture can be built. It describes the conceptual structure of systems which are to communicate.

Originator/Recipient (O/R) Address — A descriptive name for a UA which contains characteristics which help the MTS in establishing a route to the recipient.

Originator/Recipient (O/R) Name — A purely descriptive name for a UA.

Originator — A user which may be a human being or a computer process from which the MTS accepts the message for transfer to the recipient.

Originating UA — The originator's interface to the MTS. The originating UA will employ the MTS to route the message to the recipient.

Octet — This is the name used in most International Standards to indicate a byte, 'oct' implying eight, ie eight bits. As a rule then octet = byte or vice versa.

Private Management Domain (PRMD) — An MHS which is outside the control of a Service Provider, under private control for its administration and maintenance.

Physical Delivery — Physical delivery of messages originated within the MHS to users via postal, courier or other services. This service has been defined within the 1988 X.400 recommendations.

Physical Delivery Access Unit (PDAU) — The means by which an MHS user can access a physical delivery service.

Postal O/R Address — In the context of message handling a specific form of O/R address which has the characteristics of a normal postal address; in addition to this it will also identify the specific physical delivery service which is to be used (ie either postal or courier).

P1,P2,P3,P7 — different classes of protocol specified within the CCITT X.400 (MHS) standards.

Packet-Switching — A type of data network based upon the CCITT X.25 recommendation, whereby a 'virtual call' is established, but individual data 'packets' may be routed across separate physical links through the network. (British Telecom's PSS is of this type.)

PAD — Packet Assembler/Disassembler; converts data at a terminal into 'packets' (discrete quantities) for transmission over a communications line and set up and addresses calls to another PAD (or system with equivalent functionality). It permits terminals which cannot otherwise connect directly to a Packet Switched Network to access such networks.

PCI — Protocol Control Information; control information passed between peer entities to coordinate the transfer of user data. It is added to the Service Data Unit to create a Protocol Data Unit.

PDU — Protocol Data Unit; created at a given layer in the stack by taking the Service Data Unit from the layer above and adding PCI. This is the information which is passed to the peer entity.

Peer Entity — Active element within an OSI layer which corresponds to an equivalent element in the corresponding layer of a different end system.

Physical Layer — This is the first level of the OSI Reference Model, responsible for transmitting bit streams between data link entities across physical connections.

Presentation Layer — Level 6 in the model; responsible for agreement on how information is represented.

Protocol — A set of rules for the interaction of two or more parties engaged in data transmission or communication. In OSI terms interaction between two layers of the same status in different systems.

Protocol Stack — The set of OSI protocols at all seven layers required for a particular function or implemented in a particular system.

PSDN — Public Switched Data Network; CCITT term for public packet switched network.

PSE — Packet Switching Exchange; a switching computer which adheres to X.25 packet-level procedures. Used by BT to describe PSS exchanges.

PSS — Packet SwitchStream; British Telecom's packet switched public data network.

PSTN — Public Switched Telephone Network.

PTT — National postal, telephone and telegraphy organisation.

Relay — A term used for a system which performs an intermediate function in the communication between two or more END-SYSTEMS (eg a node in a public switched network). In the context of X.400 relay implies the movement of messages between MTAs.

Receipt — An X.400 message is only classified as being received when the user has actually accessed the message; delivery of the message, by the MTA, to the recipient's UA does not constitute receipt.

Recipient — The intended recipient of an X.400 message is the person or system which is specified by the O/R name or address placed on the message by the originator.

Routeing — Function within a layer to translate title or address of an entity into a path through which the entity can be reached.

SAP — Service Access Point; allows entities within adjacent layers to interact (see NSAP).

SASE — Specific Application Service Elements; those parts of the OSI Application Layer which include FTAM, JTM, VT and MOTIS.

SC — Subcommittee; within ISO, SC21 has responsibility for development of standards for OSI layers 5 to 7, SC6 for layers 1 to 4, SC18 for message handling systems.

Service — The interface between a LAYER and the next higher layer (in the same system) ie the features of that layer (and below) which are available for selection and the conditions reported.

Submission — When a message has been prepared the originating UA will submit it to the MTS for delivery to the intended recipient.

Simple Formattable Document (SFD) — Recommendation X.420 from the X.400 series contains a specification for defining the attributes of a document in an unambiguous manner. SFD, as it is known, was devised to allow the interchange of documents which can be described as formattable but not yet formatted.

Submission Delivery Entity (SDE) — An entity which is co-located with a remote UA, but residing in the MTL. It provides the remote UA with interactive access to the facilities of the MT-service.

SDE Protocol P3 — The interactive protocol which is used between an SDE and its MTA.

Session Layer — Fifth layer in the model, responsible for managing and coordinating the dialogue between end systems.

SNA — IBM's proprietary Systems Network Architecture, which is layered but at present has only some architectural similarity to OSI.

SPAG — Standards Promotion and Application Group — a consortium of European suppliers developing functional standards.

TC — Technical Committee; within ISO, TC97 has responsibility for SC6, SC18 and SC21, which are the primary subcommittees developing OSI standards.

Teletex — An international service for document interchange, which provides rapid exchange of text via the telephone network and other public data networks. Unlike telex, teletex is a method rather than a specific network or system (CCITT F.200, T.60, T.61 and T.62).

Trade Data Element Directory (TDED — ISO 7372) — A list of the standardised data elements to be used during EDI transactions. Soon to be updated and expanded to encompass the new applications intended for the EDIFACT (ISO 9735) EDI syntax rules.

TOP — Technical and Office Protocols; a set of functional standards designed for the office environment, initiated by Boeing in the US.

Transport Classed — The method by which the options of the Transport Layer are grouped into five subsets.

Transport Layer — Fourth level of the Reference Model, charged with guaranteeing end-to-end communication between end systems.

Triple X — The CCITT recommendations X.3, X.28 and X.29 — which jointly define standards for asynchronous terminals to access a mainframe (or X.25 packet terminal) via a PAD.

User — The functional object (eg a human user, computer process) which is using the facilities of the MHS.

User Agent (UA) — A UA provides the user with an interface to the facilities of the MTS. It may also provide local facilities such as document filing/retrieval and editing functionality but these are outside the realm of X.400.

User Agent Layer (UAL) — This like the MTL is strictly a 1984 X.400 concept which does not agree with current OSI Application Layer ideas. The UAL is the upper of the two sub-layers within the Application Layer which provides the user services such as IPMS.

Virtual Circuit — A logical transmission path through an X.25 packet switched network established by the exchange of set-up messages between two DTEs. The circuit may use more than one physical circuit, or share a physical circuit with other virtual circuits.

WAN — A Wide-Area Network; makes use of communications facilities which can carry data to remote sites; could be a public data network (PDN) such as BT's PSS or a private network.

X.3, X.28, X.29 — The set of Triple-X protocols.

X.21 — The CCITT recommendation defining interfaces for synchronous transmission over Public Data Networks (CIRCUIT-SWITCHED networks).

X.25 — The CCITT recommendation defining interfaces to packet-mode terminals on packet-switched networks, as used by British Telecom's PSS and many other national and private networks.

X.25 (1980), X.25 (1984) — The variants of X.25 agreed by CCITT at its plenary meetings in 1980 and 1984, respectively. The 1980 version is a subset of the 1984 version.

X.400 — The CCITT series of Message Handling Service recommendations.

Appendix B
ISO and CCITT Communications Standards

INTRODUCTION

This appendix attempts to list some of the areas of OSI standardisation being undertaken by both ISO and the CCITT. It does not constitute a definitive list, but rather an attempt to illustrate the variety and complexity of the available standards.

CCITT Recommendation	ISO/IEC Standard or Technical Report
	Architectural Models
X.200	ISO 7498, OSI Basic Reference Model (1984)
	ISO 7498/AD 1, Connectionless Mode Transmission (1987)
	ISO 7498/DAD 2, Multi-peer Transmission
	ISO 7498/2, Security Architecture
	ISO 7498/3, Naming and Addressing
	DIS 7498/4, Management Framework
	DP 9545, Application Layer Structure
	Descriptive Conventions
X.208	ISO 8824, Specification of Abstract Syntax Notation One (ASN.1) (1987)
	ISO 8824/AD 1, Specification of Abstract Syn-

CCITT Recommendation	**ISO/IEC Standard or Technical Report**
	tax Notation One (ASN.1) – Addendum 1:Covering Real, Subtypes, etc
X.209	ISO 8825, Specification of the Basic Encoding Rules for Abstract Syntax Notation One (ASN.1) (1987)
	ISO 8825/AD 1, Specification of the Basic Encoding Rules for Abstract Syntax Notation One (ASN.1) – Addendum 1: Covering Real, etc
X.210	ISO TR 8509, OSI Service Conventions (1987)
X.407	ISO 10021/3, MOTIS – Abstract Service Definition Conventions

Conformance

X.290	DP 9646/1, OSI Conformance Testing Methodology and Framework – Part 1, General Concepts
	DP 9646/2, OSI Conformance Testing Methodology and Framework – Part 2, Abstract Test Suite Specification
	DP 9646/3, OSI Conformance Testing Methodology and Framework – Part 3, The Tree and Tabular Combined Notation
	DP 9646/4, OSI Conformance Testing Methodology and Framework – Part 4, Test Realisation
	DP 9646/5, OSI Conformance Testing Methodology and Framework – Part 5, Requirements on Test Laboratories and their Clients in the Conformance Assessment Process

Common Application Service Elements

| X.217 | ISO 8649, Service Definition for Association Control Service Element (ACSE) |

CCITT Recommendation	**ISO/IEC Standard or Technical Report**
X.227	ISO 8650, Protocol Specification for ACSE
	DIS 9804, Commitment, Concurrency and Recovery (CCR)
	DIS 9805, Protocol Specification for CCR
X.218	ISO 9066/1, Reliable Transfer – Part1, Model and Service Definition
X.228	ISO 9066/2, Reliable Transfer – Part 2, Protocol Specification
X.219	ISO 9072/1, Remote Operations – Part 1, Model, Notation and Service Definition
X.229	ISO 9072/2, Remote Operations – Part 2, Protocol Specification

Peer-to-Peer Applications

ISO 8571/1, File Transfer, Access and Management – Part 1, General Description

ISO 8571/2, File Transfer, Access and Management – Part 2, Virtual Filestore Definition

ISO 8571/3, File Transfer, Access and Management – Part 3, File Service Definition

ISO 8571/4, File Transfer, Access and Management – Part 4, File Protocol Specification

DP 8571/5, File Transfer, Access and Management – Part 5, Protocol Implementation Conformance Statement (PICS Proforma)

DIS 9040, Virtual Terminal Basic Class Service

DIS 9041, Virtual Terminal Basic Class Protocol

DP 9579, Remote Database Access (RDA)

CCITT Recommendation	**ISO/IEC Standard or Technical Report**
	Multi-party Applications
	DP 10026/1, Distributed Transaction Processing – Part 1, Model
	DP 10026/2, Distributed Transaction Processing – Part 2, Service Definition
	DP 10026/3, Distributed Transaction Processing – Part 3, Protocol Specification
	DIS 8831, Job Transfer and Manipulation (JTM) Concepts and Services
	DIS 8832, Job Transfer and Manipulation (JTM), Specification of the Basic Class Protocol for JTM
	Distributed Applications
X.500	ISO 9594/1, The Directory – Overview of Concepts, Models and Services
X.501	ISO 9594/2, The Directory – Models
X.511	ISO 9594/3, The Directory – Abstract Service Definition
X.518	ISO 9594/4, The Directory – Procedures for Distributed Operation
X.519	ISO 9594/5, The Directory – Protocol Specification
X.520	ISO 9594/6, The Directory – Selected Attribute Types
X.521	ISO 9594/7, The Directory – Selected Object Classes
X.509	ISO 9594/7, The Directory – Authentication Framework
	DP 9595/2, Management Information Service

CCITT Recommendation	**ISO/IEC Standard or Technical Report**
	Definition – Part 2, Common Management Information Service (CMIS) Definition
	DP 9596/2, Management Information Protocol Specifications – Part 2, Common Management Information Protocol Specification
X.400	ISO 10021/1, Message Oriented Text Interchange System (MOTIS) – Systems and Service Overview
X.402	ISO 10021/2, MOTIS – Overall Architecture
X.407	ISO 10021/3, MOTIS – Abstract Service Definition Conventions
X.411	ISO 10021/4, MOTIS – Message Transfer System: Abstract Service Definition and Procedures
X.413	ISO 10021/5, MOTIS – Message Store: Abstract Service Definition
X.419	ISO 10021/6, MOTIS – Protocol Specification
X.420	ISO 10021/7, MOTIS – Interpersonal Messaging Systems
	DP 10031/1, Distributed Office Applications Model (DOAM) – Part 1, General Model
	DP 10031/2, Distributed Office Application Model (DOAM) – Part 2, Referenced Data Transfer
	Presentation Layer
	DIS 8822, Connection-oriented Presentation Service
	DIS 8823, Connection-oriented Presentation Protocol

CCITT Recommendation	**ISO/IEC Standard or Technical Report**
	Session Layer
X.215	ISO 8326, Connection-oriented Session Service
X.225	ISO 8327, Connection-oriented Session Protocol
	Transport Layer
X.214	ISO 8072, Transport Service Definition
X.224	ISO 8073, Connection-oriented Transport Protocol
	ISO 8073/ADD 1, Network Connection Management Sub-protocol
	Network Layer
	ISO 8208, X.25 Packet Level Protocol
X.213	ISO 8348, Network Service Definition
	ISO 8348/ADD 1, Connectionless Network Service (CLNS)
	ISO 8348/ADD 2, Network Layer Addressing
	ISO 8473, Connectionless Network Protocol
	DP 8648, Internal Organisation of the Network Layer
	ISO 8878, Use of X.25 to provide Connection-oriented Network Service (CONS)
	DIS 8880/n, Provision of Network Service
	DIS 8881, Use of X.25 Packet Level Protocol in LANs
	Data Link Layer
	ISO 7776, HDLC: X.25 LAPB Compatible DTE Data Link Procedure

CCITT Recommendation	**ISO/IEC Standard or Technical Report**
	DIS 8802/2, Logical Link Control for LANs

Physical Layer

X.21   DTE/DCE Physical Level Interface Characteristics

LANs (effectively include data link functions)

             ISO 8802/3, CSMA/CD Technology

             ISO 8802/3 DAD 1, CSMA/CD Reduced Geography (Cheapernet)

             ISO 8802/4, Token Bus Technology

             ISO 8802/5, Token Ring Technology

Physical to Network Layer

X.25   Access to Packet Switched Networks

Index